CLASSIC HYDRONICS

HOW TO GET THE MOST FROM THOSE OLDER HOT-WATER HEATING SYSTEMS

DAN HOLOHAN

For additional copies, contact

HeatingHelp.com

63 North Oakdale Avenue

Bethpage, New York 11714

1-800-853-8882

customerservice@heatinghelp.com

Or go to the Shop at www.HeatingHelp.com

First printing, December 2011

FOR THE DEAD MEN:

MR. CARLSON

MR. TIDD

AND OF COURSE, MR. HICKS

CONTENTS

CHAPTER ONE

A Simple Copper Elbow

Let's begin with a story because a wonderful history, and a simple beauty, weave their way through these older hot-water systems. It's a story of science, engineering, invention, and hard work. It's also a story of people, most of them unnoticed.

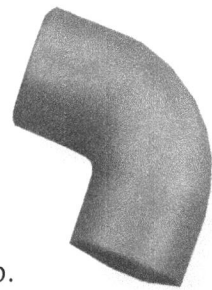

When I was 20 years old, I took a job working for a manufacturer's rep. My father was my boss, and since we were the first father-and-son combination ever to show up at this company, he was especially hard on me. He didn't want anyone to think I was being favored. We were the representatives for the Northern Indiana Brass Company at the time. NIBCO makes copper fittings, but we didn't stock any of them. The factory would ship the goods directly to the wholesalers, and my job was to keep track of who got what. I was a clerk. From nine until five I would look at factory-shipment records and make check marks on the customers' orders. I sat at an old wooden desk and shuffled paper all day long. I had a mechanical calculator that was the size of a toaster oven. It chugged back and forth like an unbalanced washing machine. Day after day, I processed paper, never really knowing what I was dealing with. I never got to meet the people who made, bought, or installed this stuff.

One day, I wandered out to the warehouse where the company kept a very small supply of copper fittings, which they used mostly for samples. I picked up a simple copper elbow and carried it back to my desk. I liked the way it blinked brown and shiny in the florescent light.

That afternoon, when the phone calls had died down, I found myself playing absentmindedly with that copper elbow. I stuck my pinkie through it and felt its smoothness. I remember wondering how they got it to be that smooth. I had never been to a factory. I could only wonder.

In the days that followed, and mostly out of boredom, I began to think more and more about that simple copper elbow. I picked it up and held it to my nose. Copper has a particular odor that's unlike anything else. It reminded me of the taste I'd get in my mouth when I ran too hard on autumn days during touch-football games. That simple copper elbow reminded me of friends who had moved away years before.

As the days went by, I started to think about where the copper came from. I imagined a mine in Chile or some other exotic place I would probably never get to visit. Chile was in the news a lot back then and I read that they had copper. I thought of the men who went down into the earth and clawed the copper from the rocks. I imagined their skin to be as brown as the copper itself. I wondered what their lives were like, if they had wives and children, and if their children would someday work in the mines too.

I began to think about the ore and how the copper got from the mines to the smelters and then (in what form?) to the factory in Elkhart, Indiana. I thought of the ships that must carry the ore northward, and the men who piloted those ships. I wondered if they got bored staring at the sea and their instruments day after day. I thought of these things as I made my check marks on the customers' orders:

One hundred #607 elbows shipped, size: half-inch.

Two hundred #611 copper-by-copper tees, size: three-quarter inch.

I checked them all, and wondered if this was how my life was to be.

When the copper got to Elkhart, it had to be unloaded and someone must be doing that hard work right now, I imagined. I tried to picture what those men looked like. How big were their arms? Did they stop on their way home and drink beer and complain about the boss while their wives waited for them with crying children? I figured these men had much in common with the men who mined the copper. I tried to imagine them meeting in some roadhouse on a gray Midwestern afternoon. Would they recognize each other?

I thought about the machines at the factory that were powerful enough to bash copper into the shape of an elbow, or a tee, or a threaded union. How much force would it take to do that? And what would it sound like? And who invented and built that machine? And what did it weigh? And who figured out all of this?

I closed my eyes and tried to imagine what it must be like to go to work in that factory, day after day after day, knowing you probably would be doing this for the rest of your life. I imagined what it might feel like to be a former, small-town football hero who has now grown a belly. He does this work every day in America, but the pounding of the machinery has replaced the cheering of the crowd, and that pounding is relentless.

And then I thought of the constant buzzing of the semi's wheels as it races along Interstate 80 on its way to New York City with the copper fittings. The driver stares as far as his head-lights will allow through a bug-splattered windshield. He smokes an unfiltered cigarette, and a country song plays on the radio. The driver stubs out the butt, and then lights another as the miles sling out from under his heavy truck. He thinks about his wife, who is as far away as next week.

And in New York City a few months later, a young man drives a forklift onto the back of another truck and unloads the fittings into his boss's warehouse. He has a date that night with a girl he will someday marry. He's not thinking about the copper, only about the girl.

The fittings sit on the wholesaler's shelf for a while and quietly gather dust. One day a heating contractor picks up a heavy cardboard box filled with fittings and a half-dozen other items and tosses it all into the back of his van. He goes from job to job, making repairs and installing new equipment, and on one Tuesday morning, he reaches into the box and comes out with a simple copper elbow. He cleans the inside of the fitting with a stiff wire brush, swabs some flux into it, and slips it onto the end of a bright copper tube. He never stops to think of how precisely it fits. He never considers what has gone into the mating of fitting and tube. People working all over the world have played a part in this precision, but none of them give much thought to this mating. They just go to work every day and do their jobs.

The contractor touches a spark to the end of his torch and watches the fire pop to life. He holds the flame to the base of the fitting and waits. The gas, which once slept deep in the earth, kisses the copper in a way that is as old as creation.

When he has finished his work, the contractor lets hot water surge through the tubing and around that simple copper elbow. He packs up his tools and walks out to his truck. He's going to take his young son to a game that night.

That evening, the homeowner sits and reads his newspaper in a warm living room. His wife watches television and his children do their homework. The man casually flips past a small article about trouble in a copper mine somewhere in a country that is too far away to concern him.

He licks the tip of his index finger and turns the page.

CHAPTER TWO

The Classics

I'm a huge fan of the new stuff. I love condensing boilers and ECM motors on pumps and fans. These things are smarter than I am and they're not afraid to tell me so. I love the sleek, sexy European radiators that I can't afford to put into my own home, but they do make me drool.

I love plastic pipe that bends around corners, and I love controls that want to talk to heating systems from hundreds of miles away while the building owner is driving 70 miles per hour on the Interstate.

I'm nuts about all of this new stuff. It's exciting and inventive and it saves fuel, but most of what I see as I wander around America is old, troubled, and in need of attention.

So that's what this book is about.

Only about 11 percent of American homes have hydronic heat, and that includes steam. The folks who manufacture boilers and radiators coined the term "hydronics" in 1946 to make the science of heating a building with water in any form sound sexy, like "electronics," a word that had been with us since 1900, but which was all the rage in those post-war years.

Most of the classic hydronic systems are older than we are, and they've gone through changes over the years, some of them, painful changes. Many of them went in before we knew as much as we know today about hot-water heating. Heck, it wasn't until the mid-'60s that we figured out where to install the circulator to keep the air from coming out of solution and stopping the flow.

The thing about America, though, is that once we put in a heating system, we tend to keep it, and keep it *as is*. Sure, we change our other appliances for fashion's sake, but I've never met anyone who wanted to change her boiler because it's not this year's model, or it's the wrong color. Even with today's "green" consciousness, most Americans would rather put solar panels on their roofs and rain barrels on their downspouts than get rid of that old boiler that crouches like a forgotten old uncle in their basement. We get excited about new cars, but a new heating system is usually just an unexpected, and certainly an unwanted, expense.

Tell me that's not true.

So let's talk about how classic-hydronic systems work, how they've changed over the years, and what we can do right now to get the most out of them without spending a fortune. My goal is to make you "see" inside those older systems, and to be able to do stuff that others say can't be done. If I can do that, I will be in a state of bliss, and you will be nodding your head and smiling at your newfound knowledge.

I'll tell you some tales along the way and I hope you enjoy the hearing as much as I enjoy the telling.

Okay, let's get started.

Think of this as a good Liberal Arts education in hot-water heating. I want you to focus on the core principles that make all hydronic systems work. Things such as temperature difference (what engineers call Delta T) and pressure difference (they say Delta P). These two concepts represent the Zen of hydronic heating, and we'll keep coming around to them as we move along our delicious, hot-water buffet. Keep those core principles on your plate at all times and you'll never go hungry.

Delta P creates flow because high pressure goes to low pressure. Always. Where there is no flow, there is no heat, so that's why Delta P is on the plate. An air problem and a Delta P problem look like identical twins, and since life can be unfair, you'll probably choose the wrong twin when you're troubleshooting. One thing we'll spend a lot of time on is the concept that, when you're bleeding a radiator, if you don't get any air, *stop* bleeding!

If you don't get any air, it *ain't* an air problem.

It's a balance problem.

By glancing back nostalgically at old gravity systems, we can begin to see the essence of Delta P, which makes *all* heating systems work, no matter what type, so we'll begin with those.

In diverter-tee systems, maddening as they can be, we see the roots of primary-secondary pumping, which arrived in the '60s, courtesy of the late, great Gil Carlson, who was my teacher.

One-pipe loop systems run the concept of Delta T up the flagpole, and show us the value of two-pipe hydronic systems. Too much baseboard means no heat at the ends of the loops because heat travels on the flow like a passenger on a train, and where there is no heat, there is no happiness.

Two-pipe systems let us deliver about the same temperature water to all the radiators, and that is its great advantage over the one-pipe system when it comes to those larger buildings. Here, too, it's all about the Delta sisters (P and T).

Classic radiant began on Long Island and I have some very good stories to tell you about those. I think you'll enjoy them.

But let's begin at the beginning and see where it all takes us. I promise to do my best to make this a fun trip. Oh, and just so you know, that simple copper elbow is sitting right here on my desk.

CHAPTER THREE

An Interview with a Dead Man

My buddy, Jim Walls, introduced me to Charles Tomlinson by sending me a copy of Mr. T's book, *Rudimentary Treatise on Warming and Ventilating*. The price marked on the cover is two shillings, which I suppose was a lot of money back in 1850, the year they published this gem. That was also the year that Mr. Tomlinson's fellow Londoner, Charles Dickens, published his novel, *David Copperfield*. In America, Nathaniel Hawthorne published, *The Scarlet Letter*, and sold 4,000 copies in the first 10 days. It was a fine year for books.

I sat with Mr. Tomlinson, who lives on in his book, for hours, and he told me much about how central heating began. I really like this guy. He's knowledgeable and always patient with me. He also has a wonderful way with words. I think you'll like him too.

He was working with heating systems at a time when folks were making it all up. There was much more concern for ventilation in those days than there was for heating, and for good reason. Because they lived with such horrible conditions, many Londoners died from what they then called carbonic-acid poisoning, or what we today call carbon-monoxide poisoning. They crowded into tiny rooms and kept the windows closed to keep out the cold. Those who could afford fuel for stoves didn't understand how to properly vent those appliances, and that's what led to so many of the deaths. It would be like you being indoors in January. It's freezing, all your windows are closed, and you're flipping burgers on your charcoal grill.

Not good.

So along comes the promise of central heating and good ventilation, and that's what Mr. Tomlinson told me about as I read his book. There was steam, and there was hot water, and the engineers were trying them both. But when it came to central heating, everything was so new in 1850, and they had so much to learn, so I sat with him in a quiet room one night and asked him these questions.

Dan: About these heating systems you're working on, Charlie. Are they for new buildings, or are you going to retrofit them into existing buildings?

Charlie: In any building where this apparatus is intended to be erected, it ought not to be introduced as an afterthought. It should be remembered that as its complete success, and its economical character, depend, in a great measure, upon due consideration of its benefits being given at the commencement of a building, so it ought, in future, to engage the primary consideration of the architect and builder.

Dan: Hmm, that could be a problem, Charlie. You ever try to sell a heating system to a builder? All they're interested in is the cheap stuff. No buyer ever looks at the heating system. And architects never give you enough room for the mechanical equipment, so you can forget about that. It's not going to happen. But tell me about these hot-water systems you're proposing. That's mostly what we're using now in the 21st Century. What operating temperatures do you have in mind for the 19th Century?

Charlie: It is of great importance to ascertain whether this apparatus is perfectly safe, for even a doubt on the subject may be fatal to its general introduction. The average temperature of the pipes is stated to be generally about 350 degrees, but a very material difference in temperature, amounting sometimes to 200 or 300 degrees, is said to occur in different parts of the apparatus, in consequence of the great resistance which the water meets within the numerous bends and angles of this small pipe.

Dan: That hot, huh? And a 300-degree temperature difference in some parts of the system? You're going to get some callbacks on that one, Charlie. And I'd be concerned about making the pipes that hot. How are you going to check the temperatures? Got a good thermometer?

Charlie: A very elegant method of ascertaining the temperature of a heated surface of iron or steel consists in filing it bright, and then noting the color of the thin film of oxide which forms thereon. In some apparatus, if that part of the pipe which is immediately above the furnace be filed bright, the iron will become a straw color, showing a temperature of about 450-degrees. In other instances, it will become purple at about 530-degrees, and a full-blue color at 560-degrees. At 600-degrees, it will be dark blue, verging on black.

Dan: You're telling temperature by the color? Yikes! Aren't you concerned that you'll be making steam inside your hot-water heating system? That won't be good for the customers. Trust me on this.

Charlie: There is always steam in some part of the apparatus. The pressure can be calculated from the temperature, and a temperature of 450-degrees will give a pressure of 420 pounds on the square inch. At 530-degrees, it will be 900 pounds, and at 560-degrees, it will be 1,150 pounds.

Dan: Blimey! And your pipes can take that?

Charlie: Although the pipes are proved at a pressure of nearly 3,000 pounds per square inch, and the force required to break a wrought-iron pipe of one-inch external and a half-inch internal diameter, requires 8,882 pound per square inch on the internal diameter, yet these calculations are taken for the cold metal. By exposing iron to long continued heat, it loses its fibrous texture, and acquires a crystalline character, whereby its tenacity and cohesive strength are greatly weakened.

Dan: Well, ASME hasn't arrived yet back in your time, and neither has OSHA, so I'll take your word for it. I'd still try to get some sort of relief valve on that system, though. Do you have those yet?

Charlie: Instead of hermetically sealing the expansion pipe, it should be furnished with a valve, so contrived as to press with a weight of 135 pounds on the square inch. This will prevent the temperature from rising above 350-degrees in any part. The pressure would then be nine atmospheres, which is a limit more than sufficient for any working apparatus where safety is of importance.

Dan: I imagine safety is of importance just about everywhere, Charlie, but what if something breaks? I mean, what if one of those pipes with the 350-degree water lets loose near where the people are? I'll bet that would register about 9.9 on the Sphincter Scale.

Charlie: If the apparatus were to burst in any part, the effects would, by no means, resemble those which accompany the explosion of a steam boiler. One of the pipes would probably crack, and the water, under high pressure, escaping in a jet, a portion of it would be instantly converted into steam, while that which remained as water would sink to 212-degrees.

Dan: Well that will certainly get everyone's attention. And this is under ordinary circumstances?

Charlie: This would have the effect of scalding water under ordinary circumstances, but the high-pressure stream would not scald because its capacity for latent heat is greatly increased by its rapid expansion on being suddenly liberated, so that instead of imparting heat, it abstracts heat from surrounding objects. The only real danger that would be likely to ensue would be from the jet of hot water, and this must, in any case be of trifling amount.

Dan: You ever hear of lawyers, Charlie?

Charlie: Solicitors.

Dan: Whatever. Can you take me back to the beginning? Do you know who was the first guy to try heating a building with hot water?

Charlie: The heating of rooms by the circulation of hot water in pipes seems to have occupied the attention of a few speculative individuals long before the attempt was actually made. The first successful trial is assigned to Sir Martin Triewald, a Swede, who resided for many years at Newcastle-on-Tyne, and about the year 1716, described a method for warming a greenhouse by hot water. The water was boiled outside the building and then conducted by a pipe into a chamber under the plant.

Dan: Greenhouse? That's small potatoes, Charlie. Who did the first big job?

Charlie: The first successful attempt on a large scale was made in France in 1777 by M. Bonnemain, in an apparatus for hatching chickens for the purpose of supplying the market of Paris.

Dan: Chickens? You're saying that we owe all of what we do today to chickens? You mean to tell me that we can trace all of hydronics back to a bunch of folks in Paris who wanted to chow down on some Buffalo wings?

Charlie: And his arrangement of the apparatus was excellent. It has been taken as a model in many subsequent methods, although the merits of the inventor have not always been acknowledged.

Dan: Yeah, but *chickens*, Charlie?

Charlie: Indeed.

CHAPTER FOUR

Classic Gravity

So it was a Frenchman, heating chicks in Paris who started hot-water heating in a way that got noticed. I'm okay with that because Paris is lovely, and a fine place to begin just about anything. And I do love those spicy wings.

Here is Jean Simon Bonnemain's hot-water system, as it appeared in 1777.

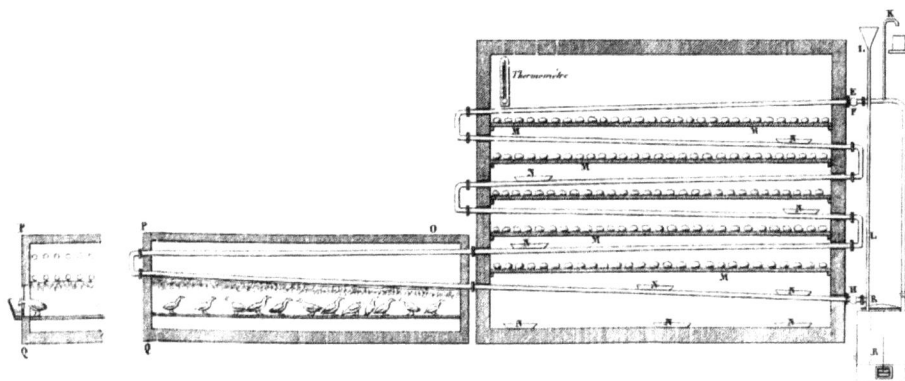

Now, this was inside a huge incubator, but we can still see all the elements of what was to come in a gravity system. We have the boiler, with a large pipe rising from it to the top of the system. The pipe is relatively large because this system has no circulator (they arrive in the 1920s). Larger pipes mean less friction to slow the rising hot water.

Near the top of the system, you'll notice a funnel. That's how you fill the system. The early gravity systems had people climbing stairs with buckets filled with water. Few places had city water mains, and filling a system, as you can imagine, took a while, especially with those large pipes.

Note, too, the overflow tank at the top. Hot water takes up more space than the cold water. That overflow tank led to what we today call the expansion tank. Expansion tanks are open to the atmosphere and give the expanded water (once heated) a place to go. The tanks that hang in the basement, above the boiler are compression tanks. They have a cushion of air that the heated and expanding water can squeeze to maintain a positive pressure throughout the system. Lots more on that later.

The heating pipes begin heating the chicks at the top and then lace their way back and forth to the bottom of the system. This would be impossible to zone, as the Soviet Russians later learned when they piped their big apartment buildings in a similar way. They're still trying to figure out what to do about that one.

The end of the heating pipe enters the boiler at the top and extends to the bottom, where the cooler water is. This is the same way cold water enters a water heater. It uses a dip tube. Hot water rises; cold water sinks. You know that instinctively.

Jean Simon Bonnemain worked his chicken magic for 15 years and then the French Revolution arrived. He survived that madness, but his business didn't. Few could afford to eat chicken during those tough times, but I'll be thinking about this giant of the hydronic heating industry the next time I'm munching McNuggets.

You should too.

Let's try it in a people house

Classic gravity heating is both the simplest and the most complicated system of all. It's simple because it has few moving parts. The water is the only true moving part. Hot water rises; cold water sinks; it's that simple. But at the same time, it's complicated because the pipes have to be just the right size for the amount of water that's flowing, and they have to have the perfect pitch. The fitter has to make sure to remove the burrs from the pipes he cuts, otherwise there could be too much friction presented to the flow. This system is so delicate that a burr can screw you up. Imagine that while you take a look at the drawing on the next page.

Oh, and notice how the boiler has two outlets, and these pipes are large. The goal is to reduce the resistance to flow so the heated water can rise. Two pipes (and big ones) allow for more flow, and that's why you'll see these multiple tappings on those classic systems.

You're also limited to a building no taller than three stories because of one of the Delta sisters, in this case, T. The hot water rises and gives up its heat to the radiators. Beyond three stories, the hot water pretty much stops rising because the water has lost much of its heat. These systems work by the difference in density between the hot water and the cold water. It's like a helium balloon. It will go just so high and then stop rising.

So here's a good question for you, as long as we're talking about helium balloons. If you were rising from the boiler, which floor would you head for first?

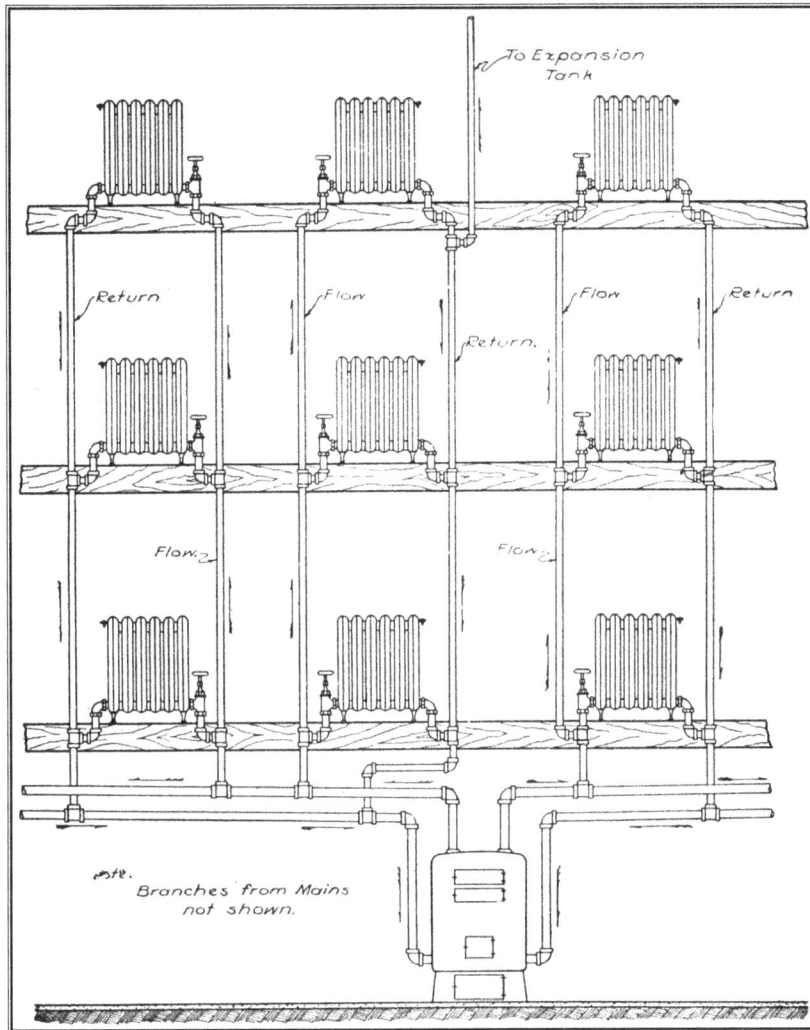

To Expansion Tank

Return   Flow   Flow   Return

Return.

Flow   Flow

Note.
Branches from Mains
not shown.

Would you go to the first floor, the second floor, or the third floor?

Think it over. I'll wait.

Did you say the first floor? Well, if you had a circulating pump on this system, the water *would* go to the first floor first. That's because water is lazy and will always follow the path of least resistance. With a circulator, that path of least resistance is the shortest route from the circulator's discharge back to its suction. In this case, that's the first-floor radiators. With a circulator on this job, most of the heat is going to go to the first floor.

But without a circulator, the path of least resistance is the *top* floor (think helium balloon). The water climbs the pipe, bypassing the first- and second-floor radiators. That means there are unhappy people in this house, and all because of the other Delta sister: P. There's less difference in pressure between the boiler and the top of these old systems than there is between the boiler and the first- and second-floor radiators. It's a Dead Man thing. They piped the job that way, and then they did something that they never wrote down because they weren't thinking about us in the 21st Century. Boooo.

What they did was add orifice plates to the top-floor radiators. They slipped them into the radiator supply valves' union connections. An orifice looks like this.

An old-timer once told me that his long-gone father used to smoke Prince Albert tobacco because it came in a can. He saved the cans and used a punch and a nail to make his own orifices. His homemade orifices worked, and he told his son about it, but he didn't tell me.

Which brings us to Mr. Hicks.

When F. Scott Fitzgerald wrote *The Great Gatsby*, he used as his model a couple of the palatial homes on the north shore of Long Island. Just south of where those homes sprawled is a town called Westbury, and in that town there was a small fuel-oil dealer that went by the name of Hicks-Westbury. When I worked for the manufacturers' rep, I used to call on those good folks. I was about 12 years old at the time, or at least that's how old I felt when I went to see Mr. Hicks. He grew up delivering oil to those mansions and many other homes in the area. He also serviced classic hydronic systems, many of which were of the gravity-hot-water variety.

My job with the rep at the time was to be the Contractor Boy. I had no direct sales duties. I was just on call to help any contractor who needed a second pair of eyes on any sort of problem job. My main talent (and I didn't have much else at that age) was that I knew how to use the library. This was before the Internet. I used the libraries in New York City and on Long Island, and I showed great respect to men such as Mr. Hicks. When I had a question, he would answer it, and I would then share what I learned with the contractor I was helping. It's amazing what older people will share with you if you just show them the respect they deserve.

The contractor had installed a new boiler in a big house that had gravity hot-water heat. The new boiler included a circulator, of course. When he started the new boiler, water flowed to the closest available radiators, which happened to be on the first floor of the house, and then back to the boiler. This is because water, as we've already agreed, is lazy.

The contractor figured he had an air problem up at the top floor, and that's why those radiators weren't getting hot. He was wrong, but he was also persistent, and he had already made up his mind. We get like that sometimes. We make up our minds while still in the truck, and then we set out to prove that we're right, in spite of all the evidence.

It's human nature.

The evidence, in this case, was that when the contractor bled the top floor radiators, he didn't get any air. He just got water. And as I mentioned earlier, when you don't get any air, it *ain't* an air problem.

But I hadn't learned that yet, in spite of being the Contractor Boy. I just knelt next to those top-floor radiators with that miserable contractor and helped him bleed the radiators, which contained no air. We labored on for hours, with great hope and anticipation, but it was no dice.

So I told the guy I'd get back to him and I went to visit Mr. Hicks. That grand old man sat and listened to my tale of woe and then said, "Did you move the orifices that are inside the upper-floor supply valves to the radiator supply valves on the lower floors?"

"What's an orifice?" I asked.

Mr. Hicks shook his head sadly and said, "*Boy*, you're a stupid kid."

And I was. But that's how I learned about what I'm telling you here. It wasn't an air problem; it was a balancing problem, and those two problems wear the same work clothes. It's tough to tell them apart, especially once you've decided that this is an air problem, even though it ain't.

But back to that house in the drawing.

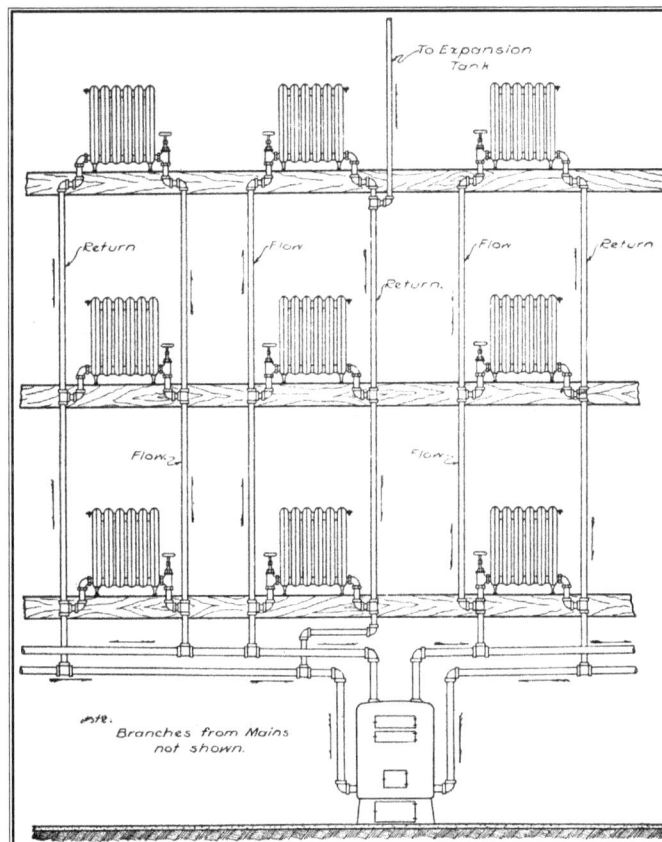

This time, I want you to notice how the pipes connect to each radiator. Notice how they enter and leave at the bottom. This also has to do with gravity flow. When the water, moving solely by the difference in density between hot water and colder water (or, as we say, gravity), enters the radiator, it looks up and thinks it's in a cathedral. It rises like a helium-filled balloon inside that spacious radiator and displaces the cooler, heavier water, causing it to fall through that lower connection on the other side of the radiator. From there, it flows back to the boiler.

This is physics at its best. It's simple and it works, but only because there's no circulator on the system.

When you change the boiler you're going to add a circulator, right?

Sure you are! What's a modern boiler without a circulator?

So you add a circulator and the water now thinks it's in an amusement park instead of a cathedral. It goes whipping around the system, giggling as it glides. It enters the radiator at the bottom and has no time to look up. It just roars through that radiator like an express train. And here's what you sense, standing on the outside: The bottom of the radiator is hot, and the top of the radiator is cold, which makes you think that this is an air problem. So you bleed the radiator. And when you bleed you don't get any air. You get only water.

Which should tell you what?

That it ain't an air problem, right?

But you've already made up your mind that it *is* an air problem, so you keep bleeding that radiator, getting only water, but as you do this, the water gets hotter. That's because you're draining the system. You're dragging hot water up from the boiler. You never get any air, but the radiator gets hot, and you see this as an indication of your sheer brilliance. You bled the radiator and it is now hot.

Hey, who can argue with that?

So you leave the job, and within a few hours, the customer calls you back because the radiator is once again as cold as a tomb. So you return and bleed the radiator again. And again. After a while, the customer has a coffee mug with your name on it.

And that's never a good thing.

Here's something else Mr. Hicks taught me: When you want more heat from a radiator that's connected bottom-to-bottom, and the system used to run on gravity, but now has a circulator, close the supply valve on that radiator and then open it just a bit. The water, being driven by the circulator, will experience the pressure drop (Delta P) of that nearly closed radiator valve and it will be like hitting a wall. The water will slow as it enters the radiator. Then it will look up and realize that it is once again in a cathedral, not an amusement park. And it will rise.

Try it because it works, and it will make Mr. Hicks smile down upon you from heaven.

When you're sizing that new boiler

Boiler manufacturers rate their boilers and they use a bunch of terms to do this. Let me take a few minutes here to explain (or at least try to explain) what that's all about.

When you look in a boiler manufacturer's catalog you'll see that there are several types of ratings for each boiler. There's the Input rating, the Gross Output rating, which they call the DOE Heating Capacity nowadays (DOE is the U.S. Department of Energy), and then there's the Net Output rating. You pick one or another to size a boiler. But are you sure you picked the right one?

To make things even more confusing, some of the ratings are shown as Btuh (British Thermal Units per Hour), while others are listed as Square Feet Equivalent Direct Radiation, or EDR for short (and here you'll find different numbers for water and steam boilers). Then we have this other column for Gallons Per Hour, which applies to fuel oil, and another for Therms, which applies to gas.

Okay, here's what's going on. You have three basic columns. First there's Input. That's where you'll probably find the ratings in Gallons Per Hour or Therms because this column has to do with fire. What you're seeing here is the amount of heat that the fire is putting into the boiler. You PUT the fire IN and that's why they call it Input.

But, not all the heat that enters the boiler winds up in the water. Some of that heat goes up the chimney and is lost forever. There's also some more heat lost through the boiler's jacket, but this is one of those vague areas because if the boiler is inside the house, can we really say that the jacket losses are gone for good? And there are some boiler manufacturers who will tell you that their jacket insulation is so fabulous that British thermal units hardly ever chose to leave that way, but whatever.

Next, we get to Gross Output (or DOE Heating Capacity). Gross is what's left over after the boiler has suffered the loss of heat up the chimney and through the jacket. Now, this term can be a bit confusing because Gross usually implies that you're dealing with the whole enchilada, as in Gross Income (which means *before* taxes, right?). But in the world of hydronics, Gross means, "what's left over" instead of "what you start with." Or to put it another way, Gross means "after taxes." "Taxes," in this case, being the price you pay when you send some heat up the chimney and through the boiler jacket. Just remember this: Gross Output is the amount of heat that rides on the water that's flowing out of the boiler. It's the heat that's available to the whole system (and maybe that's why they call it Gross).

Ready for the next factoid? Here goes: The difference between the Input and the Gross Output represents the combustion efficiency of the boiler. For instance, if a boiler has an Input of 200,000 Btuh and a Gross Output of 160,000 Btuh, that boiler would be running at 80% combustion efficiency. It's not hard to figure this out. Just divide the larger number into the smaller number and then multiply the result by 100 to get a percentage.

And this brings us to Net Output. "Net" is what you're left with after taxes, right? What's important to know here is that Net Output is always going to be less than the Gross Output because there are two things going on out there in the system.

First, we have the piping losses. By this, I mean that it's going to take a certain amount of heat to bring the pipes from room temperature up to the temperature of the water that's flowing through the pipes. And this is where things can get a bit sketchy. Are those pipes inside the living space? And if they are, is that heat really lost? Are those pipes insulated? And if they are, how insulated are they? Is there 180°F water flowing through that pipe out to baseboard convectors? Or is there 110°F water running through that pipe to a radiant panel? All of these variables should make a difference, shouldn't they? But the boiler manufacturer doesn't know how you're using the boiler, or on what sort of system you'll be installing it. They have *no* idea.

And consider this. Part of this piping-loss business has nothing to do with the heat loss of the water into the air. It has to do with raising the temperature of hundreds (or perhaps thousands) of pounds of steel and copper from room temperature to the temperature of the water that's flowing through the pipes. This isn't about heat loss. This is about heating a *lot* of metal. So, are you using a cold-start boiler (which means you have metal to heat every time the boiler starts), or is the boiler operating on an outdoor-air reset control (where there would be less of a temperature rise each time it starts)? All of this changes depending on how cold it is outside, of course. The boiler will run longer on the colder days, and that will keep the metal hotter for a longer time.

Still reading?

Thanks!

Okay, that brings us to the part that fascinates me the most - the Pickup Factor. You should know that this has nothing to do with the pipes; it has to do with the amount of water that's inside the pipes and the radiators when the boiler first starts. The boiler has to heat all that water from room temperature to, let's say, 180 degrees. That's going to take some doing if there's a lot of water in the system. Consider a classic, gravity-hot-water system. We could have six- or eight-inch-diameter pipe in some of those old beauties, and that represents a bodacious amount of water. You could also have radiators that are as big on the inside as a room at the Holiday Inn.

If this gives you pause, know that you are not the first to show concern. In 1940, the factor that manufacturers used to designate the difference between a boiler's Net Output and its Gross Output was 1.56. In other words, back in those days, you figured out the building's heat loss, and you selected your radiators to overcome that heat loss. You then took that installed radiation load in Btuh (or Square Feet EDR) and multiplied it by a factor of 1.56 to get the boiler's Gross Output rating.

That's some increase, isn't it?

You know why they did it this way? First, because they weren't sure what they were doing (all of this was so new to them). And second, because they were dealing with a lot of gravity systems that contained a bodacious amount of water.

In 1945, the war was over, building was booming, and the folks who make boilers figured that they were probably being too conservative with that Pickup Factor, so they reduced it from 1.56 to 1.33. I like to think they realized this while trying to install a boiler that was the size of Rhode Island into Mrs. Murphy's two-bedroom bungalow.

Then, in 1967, they further reduced the Pickup Factor from 1.33 to 1.15, but only for hot-water boilers. They let the factor for steam boiler sizing remain at 1.33 because steam piping is usually larger than hot-water piping.

And that brings us to modern systems. The pipes are well insulated and they're all inside the building. The system is running on an outdoor-air reset control, so the water temperature is pretty tepid most of the time. It could be a radiant system that's running on water cool enough to shower under. We're using an indirect, domestic-hot-water heater. That pretty much removes the domestic load from the equation because we're also using a priority relay that plays with time. We've got a half-dozen heating zones on this job because people are rarely in all the rooms at the same time. Heck, I once met a guy who regularly removes 10 percent of the total heating load for each zone he installs. He does this up to 30-percent total reduction from the full load and he has never once gotten into a jam.

And the piping we're using nowadays is pretty small, isn't it? So when it comes to snazzy, modern hydronic systems, maybe we need to look at that Piping and Pickup Factor again. It's something to think about, but we also need to think about it when it comes to the old gravity systems. If it's a pre-war boiler that you're replacing (and isn't it crazy that so many of those are still out there?) someone sized that beast with a Piping and Pickup Factor of 1.56. If that boiler went in after World War II, someone sized it with a factor of 1.33. You're on the job now, doing a proper heat-loss calculation, and you decide to select the new boiler from the boiler manufacturer's Net column, which gives you a built-in Piping and Pickup Factor of 1.15.

But that isn't enough for a classic hydronic system. Your new boiler, sized the modern way, is too small because it contains all that pre-war water in the pipes and the radiators. If you pipe this boiler as you would a boiler serving copper fin-tube baseboard, it will run all day and on into the night, probably burning more fuel than the old boiler did.

This is why boiler manufacturers show this diagram in their installation-and-operating booklets.

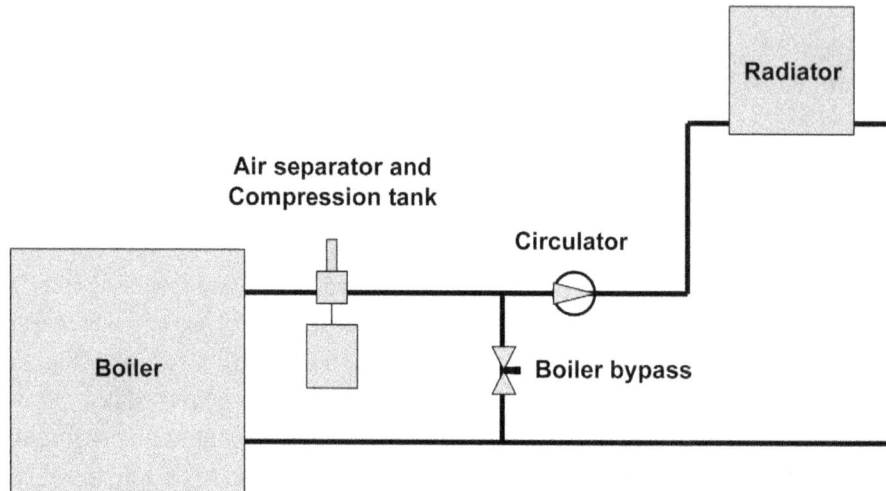

We call it a boiler bypass line, and its job is to take most of the water that's flowing around the system and have it bypass the boiler. This tricks the modern boiler into thinking it's working with a modern hydronic system that contains much less water than an old gravity system contains. The boiler gets only a portion of the flowing water to heat with each pass and the boiler is then able to fire up to its limit and shut off, saving lots of fuel.

It pays to read the manual.

And while we're on this subject, I'd like to give a big shout-out to these modern boilers that condense the flue gases. Condensing boilers love those old gravity systems because all that water allows the flue gases to condense even more efficiently. They also solve the possible problem of an ancient, oversized chimney. You don't need a chimney with a condensing boiler.

They do cost more than conventional boilers, but they're worth a look when you're working with older, high-volume systems.

Expansion tanks

So the boiler heats the water, it expands, gets less dense and rises. The colder, more-dense water falls. It makes for a lovely, hydronic Ferris wheel. But as the water expands, it needs a place to go, so the Dead Men who installed these systems used expansion tanks, and they usually put them up in the attic. Here's a picture of one on the right.

They put the tank in the attic because that was the high point of the system. The tank sometimes had a float-operated valve, similar to the one you'll find in a toilet tank, and that made the filling automatic. If there wasn't a float-operated valve, you'd fill the system by hand, turning a city-water valve, and watching for water to overflow through that pipe that goes through the roof. Then you'd have to drain a bit of water to bring the level down to about the midpoint of the tank.

This hand-filling business was tedious and it led to the development of the altitude gauge.

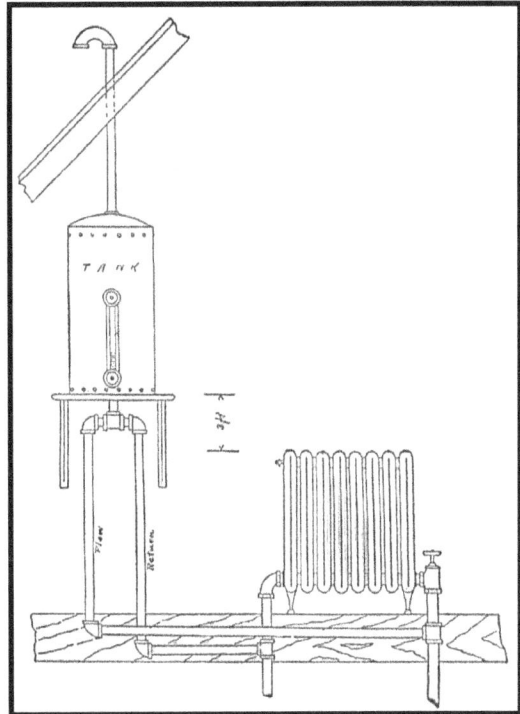

Expansion tank in an attic

You don't see those too often nowadays. They were standard pressure gauges that had an additional needle (usually a red one) that you could move with a small screwdriver, or by hand. One pound per square inch of water pressure will lift water 2.31 feet. The altitude scale on the gauge showed feet of vertical rise that corresponded to that conversion. An installer would measure the lift from where the fill valve was to the center of the expansion tank and set the red needle on the gauge to that altitude. Then he'd open the valve, allowing water into the system.

As the water stacked, the static weight of the water appeared on the pressure gauge. When the black needle reached the red needle, the installer knew there was water in tank.

Oh, and by the way, he probably wasn't working alone. Keep in mind that most of those old radiators connected bottom-to-bottom, so he needed some help because he wanted the radiators to fill with water as the water rose into the attic tank. The installer's helper went to the first floor and opened all the vents on all the radiators. When water started to squirt from those first-floor vents, the helper would shout down to the installer to shut

Altitude gauge

the valve, and then he ran around like a chimpanzee, shutting the vents before the water could cause any damage. He did the same on the second- and third floors. It was the only way to fill one of those old systems because if you tried to vent the radiators after the water reached the tank, the water would just fall out of the tank and you'd have to spend more time filling.

Here's something else Mr. Hicks taught me: Some of those old houses had no expansion tank. The installer would leave air in the top of each radiator and that combined collection of air acted as the expansion tank for the whole system. I learned this the hard way after spending hours with a miserable contractor, hunting through an old house looking for the tank that was never there. We had, of course, bled the radiators and the boiler's relief valve was popping.

Mr. Hicks reminded me of what I already knew when he commented, once again, *"Boy, you're a stupid kid."*

Guilty as charged.

Baby, it's cold inside!

And it's also cold in the attic, and that's where the expansion tank was, so freezing was sometimes a problem. To solve this, some installers piped a hot-water line into the attic, and had it circulating by gravity just below the tank. It looked like this:

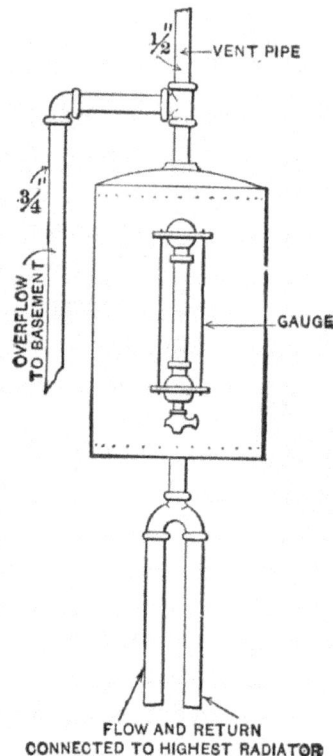

½″ ← VENT PIPE

¾″

OVERFLOW TO BASEMENT

GAUGE

FLOW AND RETURN
CONNECTED TO HIGHEST RADIATOR

Chapter Four: Classic Gravity

The hot water that was rising to the radiators also rose into that loop just below the tank. From there, it would rise up that single pipe and enter the tank. The cold water in the tank would fall down that same pipe and join the flow back to the boiler.

This points out a neat principle of hydronics that some installers miss. You don't need two pipes to get gravity flow. Hot water and cold water can pass each other in a single pipe, especially if that pipe is large. Just think about how hot water will rise in a pot on a stove as cooler water sinks. It's the same thing here.

This is also what will make hot water back out of a boiler through a return line and heat a nearby radiator in a zone whose circulator isn't running, even if there is a flow-control valve on the supply side to that circuit. And by the way, you solve that problem with a second flow-control valve, this one on the return.

Before long, the installers started moving the tanks to the basement where they couldn't freeze. Down there, they couldn't have open tanks, of course, and that's when they started calling these things compression tanks.

Expansion tanks let water expand.

Compression tanks let air compress.

It's all in the name.

They hooked these compression tanks right into the closest hot-water-supply pipe, and that usually caused the air cushion to come out of the tank and enter the water flowing to the radiators. This, again, was the fault of gravity, and I'll tell you more about it when we get to the chapter on Air.

Oh, and before we leave these tanks, I need to mention that many of them were made of copper. They are big and shiny and heavy with that now-expensive metal, which makes them quite desirable to some contractors who like to stop at the scrap yard on their way home.

I've watched contractors explain to customers how this metal is toxic, explosive, and carcinogenic, all in one, and how it must be removed from the building immediately. I even met a contractor who traveled with a large, horseshoe magnet. He'd hold the magnet to the copper tank and explain to the customer how this metal has lost its magnetism because of its advanced years, and must quickly be removed for the safety of the family and the entire town.

He was quite a guy.

Pipe tricks

When you don't have a circulator to move the water, you have to put physics to work for you, which is part of what made these gravity systems so tricky to install. Consider a large, horizontal pipe that's reaching across a basement to feed radiators on the floors above. The heated water enters that pipe and flows quite naturally to the top of the pipe. The hot water is at the top and the cold water is at the bottom of the pipe. An engineer would call that stratification and the Dead Men used it to their advantage. They knew the water's inclination was to go to the top-floor radiators first, and they wanted the system to balance, so in addition to using those orifices in the supply valves on the top floor, they also piped the connection from the horizontal mains in the basement to each riser, based on where the riser was heading.

For instance, they would pipe the radiators on the top floor from the side of the horizontal main. Like this.

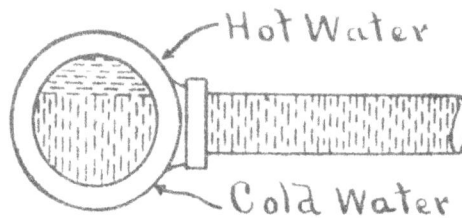

And they would pipe the radiators for the lower floors like this.

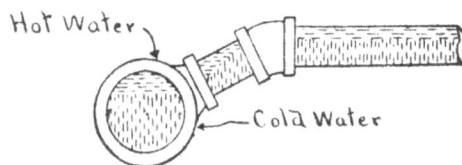

You're most likely going to be using circulators, so this doesn't mean that much to you, but what it does do for you is tell you where the pipes are going, and that sure is easier than opening the walls if you need to find out.

If it's coming off the side of the horizontal main, it's going to the top floor.

Good to know, right?

The Dead Men also used this fitting, made by Phelps.

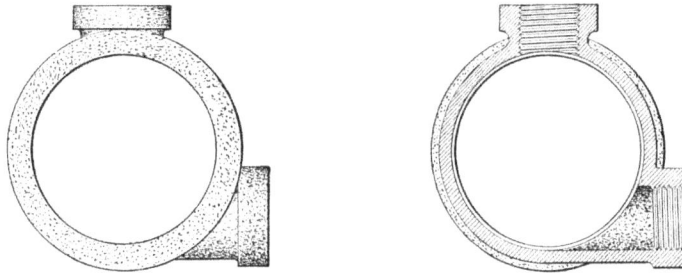

There's nothing inside that fitting except water. The hot water will stratify to the top of the fitting and flow to the radiator, while the cold water that's in the radiator will fall and enter the Phelps tee through that side connection, joining the colder water that's already there. What you have to ask is: What happens when you add a circulator to a system that has *these* fittings?

Think that radiator will heat as it should?

I don't either.

It pays to poke around a lot before you start taking things apart on these old systems. If you're not sure what you're dealing with, take a photo of it and post it on the Wall at HeatingHelp.com. One of us will probably have the manufacturer's original literature on the thing. And you usually get an answer within an hour when you post there.

Here's another example of a strange tee, but one that's not as volatile as the Phelps.

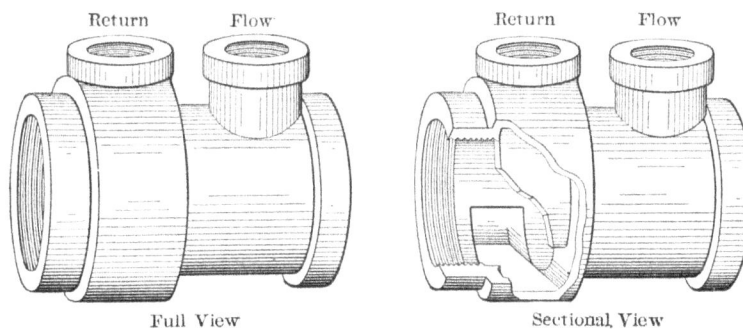

Eureka made an early version of what we now call the diverter tee. It worked by creating a drop in pressure along its run (Delta P), which encouraged the water to flow into the radiator. Here's a picture of one found on a job. The contractor who sent this kept his hands in his pockets until he learned what it was. Smart man.

Eureka fitting on the job

And then along came Mark Honeywell, who was brilliant and solved the problem of stratification on those long, horizontal runs of pipe by inventing this unusual tee.

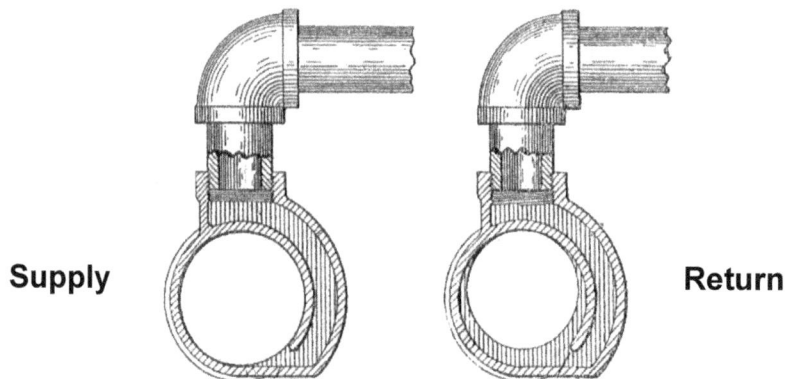

Supply   Return

The water inside the horizontal main has to be hot all the way to the bottom before it can slip under that metal wall and flow up to the radiator. This meant that all the radiators along a horizontal main would get hot water at about the same time, which helped balance the system. Wonderfully simple!

Here's drawing of the tees feeding a series of radiators.

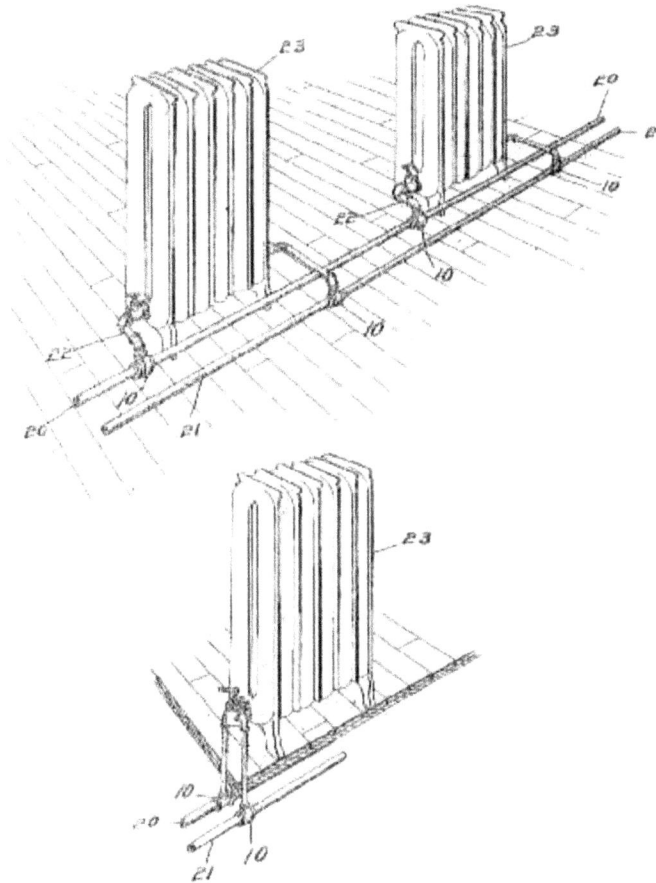

The beginning of something new

July 1916, an obituary

Oliver Schlemmer, Sr. of Cincinnati, Ohio, founder of the plumbing and heating firm, Oliver Schlemmer Co., died in his home of that city after several weeks' illness. Mr. Schlemmer was 60 years old and had lived all his life in Cincinnati. He served his apprenticeship with the heating firm of Thomas Gibson Co., and in 1879 formed a partnership with Richard Murray, under the name of the Murray-Schlemmer Co. This partnership was later dissolved and Mr. Schlemmer then formed the Oliver Schlemmer Co. and the Gibson-Schlemmer Co., the former specializing in heating work and the latter in plumbing work. Mr. Schlemmer was the inventor of the O.S. fitting used in heating work and was the first in this country to use the Spence boilers, which were a Canadian product made later by Pierce, Butler & Pierce Manufacturing Co. He was one of the charter members of the Cincinnati Master Steam and Hot Water Fitters Association. He leaves a wife and two sons, Oliver Schlemmer, Jr. chief engineer for the General Fire Extinguisher Co., Warren, Ohio, and Edmund Schlemmer, of the Office of the Supervising Architect, in Washington D.C. It is announced that the business will be continued by his brother, Clifford B. Schlemmer, who was associated with him for several years.

Not a bad life, eh? And I introduce you to Mr. Schlemmer at the *end* of that life because of the invention that he brought to the heating industry – the O.S. fitting.

At the turn of the 19th Century, central heating was new and split into two camps: those who favored steam at low pressure (what they called vapor heating), and those who preferred gravity hot water because it seemed safer. The gravity systems, up to that time, were all two-pipe systems, and the pipes had to be large to keep the frictional resistance presented to the rising hot water low. This meant that hot-water systems were more expensive than steam systems, which had large supply pipes, but small return pipes, and were generally easier to pipe.

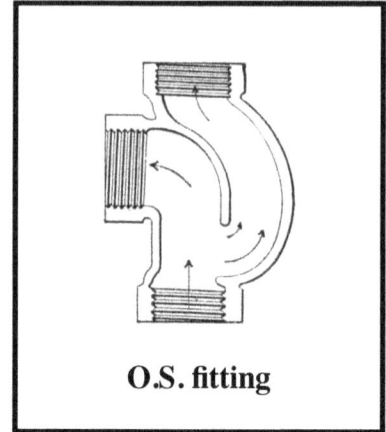

O.S. fitting

Mr. Schlemmer comes up with the O.S. fitting just as we enter the 20th Century and that introduces one-pipe hot-water heating for the first time. In the O.S. fitting, we see the roots of the diverter tee, as well as inklings of primary-secondary piping, which will follow 50 or so years later. Here's a drawing of that fitting at the top right.

This tee gave the hot-water people a better chance at getting the work when competing with the steam people. Instead of needing a separate supply and return, each radiator received its hot water from, and returned its cooler water to, just one pipe. It worked like this:

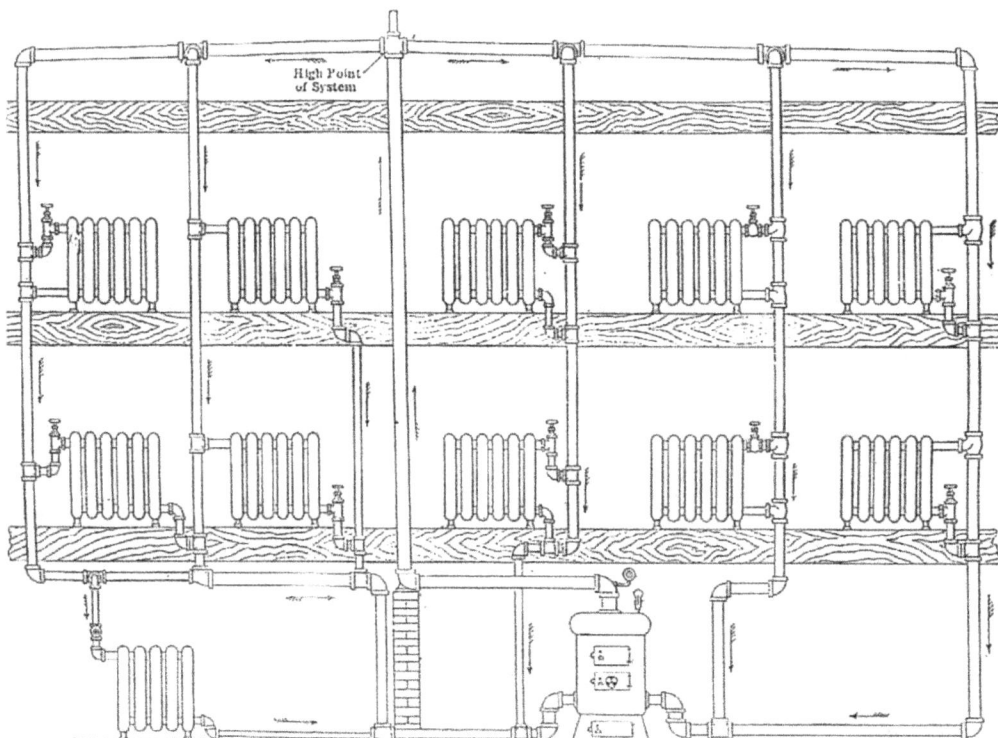

In this system, the water flows from the boiler to the top of the system, and then down to the radiators. Since the O.S. fitting has that scoop inside of it, it acts as a pressure drop to the water that's trying to stay in the pipe and flow back to the boiler. Delta P kicks in because the O.S. fittings make it easier for some of the water to flow through the radiator than it is for it to stay in the pipe. Look at the drawing and *feel* it. If you were the water, which way would you go? And you can control the heat at each radiator by throttling the supply valve at each radiator.

One-pipe hot-water heat had arrived.

Back to Mark Honeywell

Mr. Honeywell was a hot-water man, and he had some ideas of his own when it came to competing with steam. In 1902, when he was 28 years old, he founded the Honeywell Heating Specialty Company in Wabash, Indiana. He knew that steam had its advantages. It was easier and less expensive to install than gravity hot-water heat. Steam heated very quickly. The radiators were smaller than the radiators a gravity system needed. The return pipes were also smaller than the return pipes in a hot-water system, and steam doesn't know up from down. It's a gas and it will travel in any direction. It's just looking for a way out, and steamfitters accommodated the steam with properly placed air vents.

But steam also had disadvantages. The pipes were often noisy, and steam systems had a questionable reputation during the 19th Century because they blew up a bunch of buildings. At one point, steam boilers were taking down buildings at the rate of one every 36 hours.

Nice, eh?

So Mr. Honeywell set out to beat them, and he did that with a couple of brilliant inventions. One was his Unique valve, which gave him a way of connecting a hot-water radiator at just one side instead of two sides. The Unique valve opened with a quarter turn and used a baffle to direct the water into the radiator. The cooler water flowed by the hot water and entered the Unique valve on the other side of the baffle. Here's a drawing.

You control the amount of hot water going to the radiator by using the quarter-turn valve. Oh, and note the manual air vent at the top of the first radiator section. Water rises like the tide when filling a system such as this, and having the vent on the valve side of the radiator made it easier for the installer to fill the radiator if he was working alone.

Here, the valve is wide open and hot water flows from the supply pipe into the radiator, while the cooler water uses the other side of the baffle.

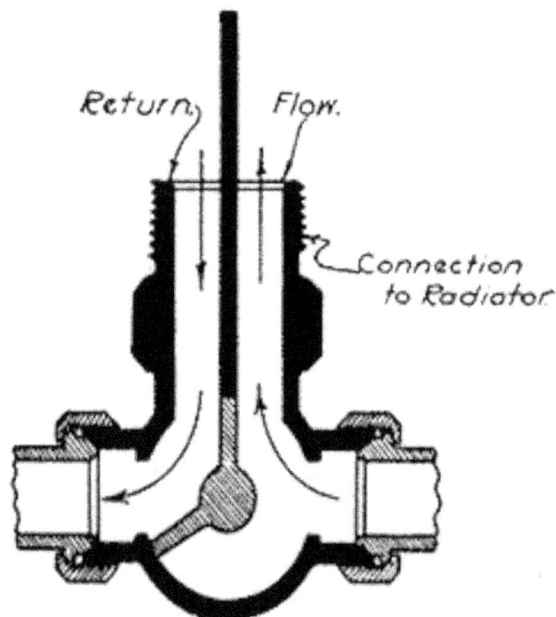

And here the valve is closed. See how the water bypasses the radiator and flows on to the next radiator, or back to the boiler? Simple, isn't it? There's very little pressure drop because of the wide-open spaces inside the Unique valve.

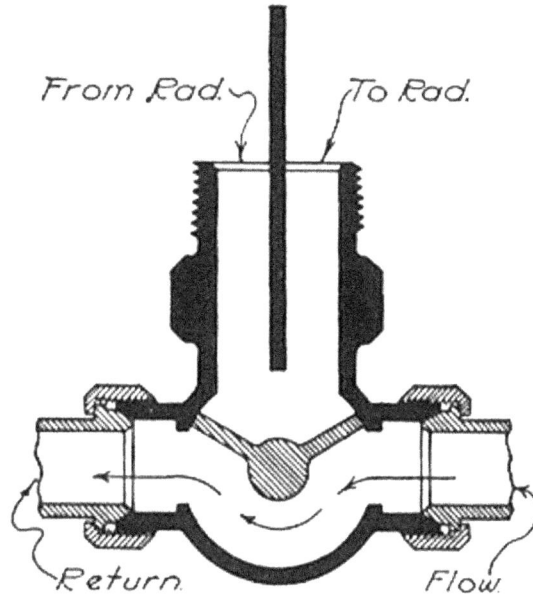

Here's a Unique valve on another radiator. One of the visitors to HeatingHelp.com posted these photos on the Wall. He had just moved into the house and was trying to figure out what he had. There's no circulator on this system.

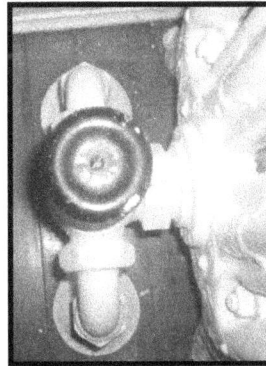

Quarter Turn Action

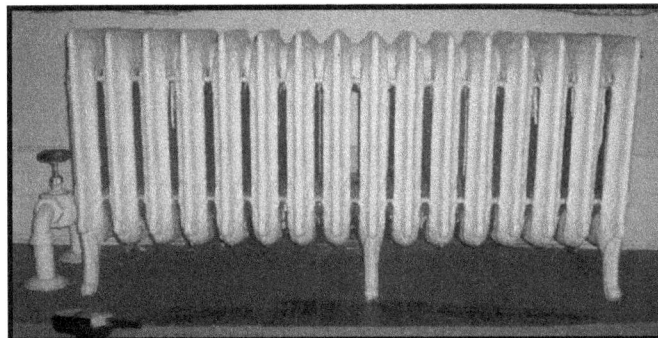

That small baffle inside the Unique valve can make water flow all the way across this long radiator. It's physics at its best. Hot water rises; cold water sinks. Even on a radiator that's this long, the Unique valve will work. This gives you a good sense of how far gravity flow will go, even on a horizontal plane. It works because the pressure drop is low.

But Mr. Honeywell faced another problem. Since gravity systems are open to the atmosphere (they have those open expansion tanks in the attic), the system designers had to limit themselves to a high-temperature of 180 degrees. Make the water any hotter than that and it might start boiling. Keep in mind that they're working at a time when the only control on the boiler was a damper that regulated the flow of air into the fire, and those weren't very accurate.

The steam people, on the other hand, were working with temperatures that *started* at 212 degrees and went up from there. That meant their radiators were smaller and less expensive than the hot-water man's radiators.

So Mr. Honeywell set out to invent a device that would allow a hot-water system to work at temperatures as hot, or even hotter, than the steam systems of the time, but *still* be open to atmosphere, which made them safe.

He came up with the Heat Generator, and here it is in the photo on the top right.

Which brings me to a story

Dave Nelsen and I used to spend a lot of time in the basements of Long Island's Gold Coast. Dave worked for Kurz Oil, and they specialized in those Great Gatsby homes. Dave and I learned together by being ridiculously curious, poking around a lot, and, when all else failed, by looking things up. Dave died way too young, and I wish we had just one more old house to visit.

He called me one day because he had come across something on a job that he had never seen. "It's made of iron, looks like one of those old-fashioned, floor-standing ashtrays, and cast into its top are the words, Honeywell Heat Generator Number One. I know they had to start somewhere with all those numbers, and it looks like I've found where they started. I got Numero Uno!"

"What are you doing there?" I asked.

"We're changing the boiler," he said.

"I'll be right over."

When I got there, Dave had disconnected the Heat Generator. His guys were tearing down the boiler, which was the size of my kitchen. "It's over here," Dave said, walking to the odd Honeywell device and picking it up. He shook it. "There's something inside," he said. He shook it again. "It's probably sludge. It's been on this system for all these years. I'll bet it's some sort of dirt collector."

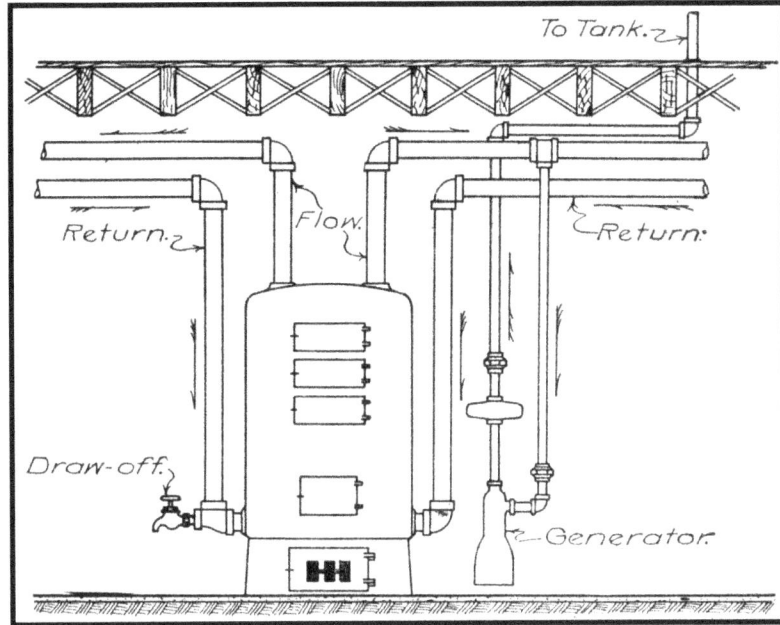

"Why would they call it a Heat Generator if all it does is collect dirt," I said. (Mr. Hicks, up in heaven, was shaking his head at this point and mumbling. *Stupid* kid.).

"Maybe it makes the heat generate faster because it gets the dirt out of way," Dave said. "I don't know. I've never seen one of these before." He shrugged and tipped over ol' Number One. It looked like he was pouring a pitcher of beer, but beer didn't flow out.

Mercury did.

Yep, pure mercury, and about a pint of it. It flowed like a liquid lawsuit down onto the floor and did what mercury will do when dropped from waist high.

I looked at Dave and Dave looked at me. You could have fit a softball in either of our mouths.

Mr. Hicks snickered. I could hear him.

Stupid kids.

So here's the deal with this thing. Steam was hotter and faster than hot water, but *way* more dangerous, so how do you make the hot water move faster so that classic system can be more efficient? Remember, they won't see a circulator until 1929, and that's way down the road.

So what do you do?

If you make the water hotter, it will flow faster, sure, but you're limited to 180 degrees because of the open expansion tank. Can you increase the temperature without making it dangerous? If you can do that, the system will heat quickly and you'll be able to use smaller pipes and smaller radiators, and you'll be able to better compete against the steam people.

You can do it with the Heat Generator because it has a pipe inside of a pipe, and both of those pipes dip into a pot of mercury. Like this:

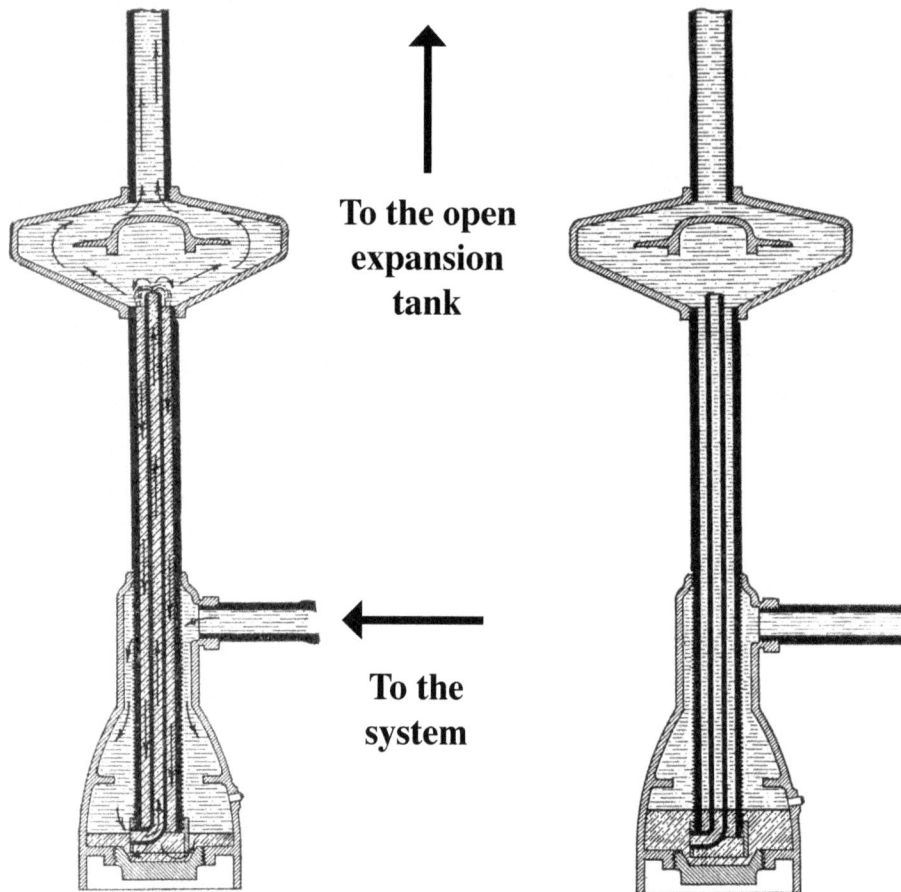

To the open expansion tank

To the system

When the water in the system heats and expands, it will push down on the mercury, which will act like a spring. As the water presses down, the mercury will rise up that pipe. You can separate the water in the piping and radiators from the water in the open expansion tank by using the mercury. Here it is in the system. Notice where the pipes go.

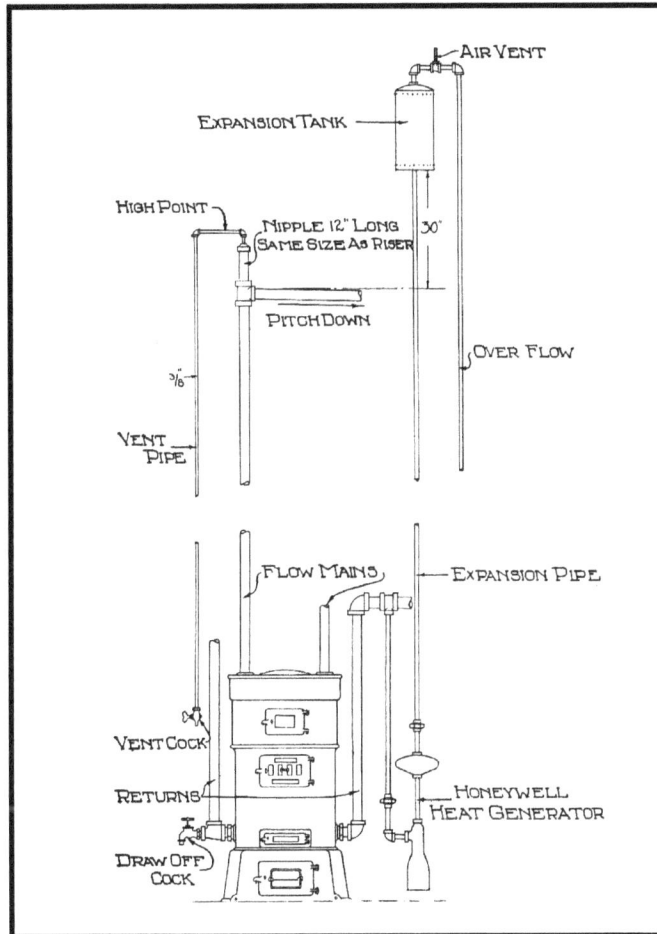

The Heat Generator gives us a way to make the water as hot as 250 degrees Fahrenheit and still have it in its liquid form because it's under pressure within the pipes and radiators, just as it would be in a modern system, but the tank in the attic is still open to the atmosphere.

Notice how the mercury rises and falls within that pipe as the system water heats and cools. If the pressure gets too high, the excess water will join the mercury inside that internal pipe and relieve itself in the separation chamber near the top of the Heat Generator. From there, the excess water will rise safely to the open tank in the attic.

It's brilliant, and in a way, it was our first "circulator" because it made the water move more quickly than gravity could make it move. It brought Honeywell to the forefront of the hot-water-heating business.

There are still plenty of Heat Generators out there, so if you come across one, don't do what Dave and I did. Drain the system, but dispose of the mercury properly. It's carcinogenic, not to be messed with, and now you know all about it.

Time passed, and frozen attic tanks gave way to closed compression tanks in the basement, but the Dead Men kept using those Heat Generators because they still didn't have circulators. Here's one of those systems.

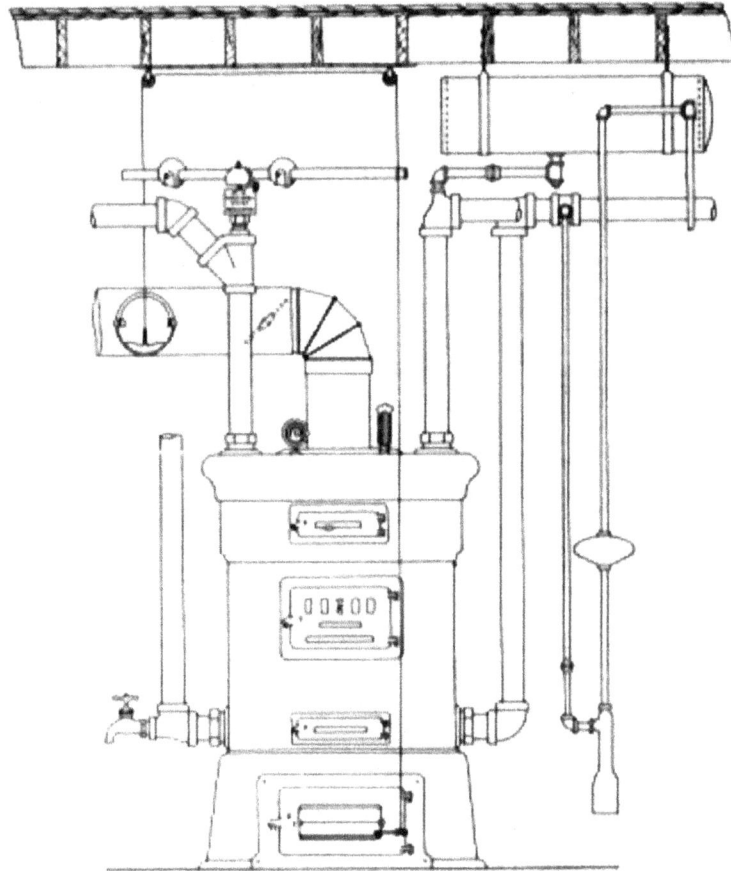

This drawing is from 1915. That's about when the tanks made their move downstairs, but check out how this installer connected the tank to the system. You see how it's right off the supply pipe? That's going to cause the tank to lose its air cushion and waterlog. Later on, I'll show you how to keep that from happening.

A story with great gravity

Back when our business was new, and before I had written any books, I was making my living doing seminars and consulting with building owners about their heating problems. I would go to their home or apartment building, look for the problem, and write a report. I didn't do any of the actual work; I just told them what was wrong and what they should do from there.

One day, the new owner of a house that looked like the one in the picture (but isn't) got in touch with me and told me of his frustration. Every 10 days or so, a fuel-oil truck that was the size of Saudi Arabia grumbled up his street. The driver would stop, pull a long hose that was as thick as your thigh up to their oil tank's fill pipe, plug in, and whistle as the oil gushed from the truck to the tank. The new homeowner hated this man.

So he hired me to see what I had to say about him saving some fuel dollars. He was open to any suggestion. Should he stick with oil, or should he switch to gas? Should he keep the boilers he had, or get new ones? Should he have the whole system repiped? Should he burn down the house? What to do?

I met him at his house, which was slightly smaller than Versailles. The oil truck was there. He asked me what I wanted to do first and I suggested we take a tour of the place, mostly because I'm nosey and never pass on an opportunity to see a grand old house.

We went from room to room and I ogled and gawked. The place had gravity hot-water heat and any of the cast-iron radiators could have held a cruise ship in place. It was a classic.

"Can we go to the basement?" I asked.

"Sure," he said. "Wait 'til you see these boilers."

"You have more than one?"

"There are two," he said. "And they're big."

"And they're connected to that truck," I stated.

"Yes," he said. "There's that."

We got to the basement, where my legs suddenly stopped working because this place had horizontal mains made from eight-inch, *screwed* pipe. You ever see eight-inch screwed pipe? Trust me; it's bigger than eight inches. I stood there with my mouth open, thinking that there

---

was once a day in America when someone lifted those bulky hunks of iron and caught a thread. Who was that guy? What sort of wrenches did he own? Where the heck did he stand?

I figured it was either Popeye or Bluto who did this job.

I was getting all sentimental and sloppy over my jaunt through time and the Saturday cartoons when the homeowner gently reminded me about those boilers, so we walked a while through the basement and that's when I came upon the two Ideal Redflash boilers. Once again, my legs stopped working. These two had once burned coal, but were now connected to the monster truck out in the street. Each had a rating of 500,000 Btuh, and both were firing.

But then, firing doesn't quite do what was going on justice. These beasts, each of which was the size of a minivan, now had tons of sand where the coal grates used to be. The burners, each bigger than the Fourth of July, hung from the doors and belched fire deeply into the bowels of those boilers. We could have toasted rye bread on the jackets.

I shut off the burners and carefully opened the door of one of the beasts. You could yodel in this boiler and get an echo. You could cremate your spouse in this boiler and no one would ever know. You'd probably get a No. 10 smoke for a few minutes, but other than that, you'd be fine. Seriously.

"Has anyone done a heat-loss calculation on the house?" I asked the homeowner.

"Yes," he said. "The salesman from the oil company did one."

"Oh."

"Yeah, it didn't take him long," the homeowner said.

"Did he go from room to room and measure the walls and windows and whatnot?" I asked. "Did he check the attic?"

"No," the homeowner said. "He just wrote down what was on the two labels and then gave me this quote." He took a sheet of paper from his folder. The quote was for a single boiler, rated at 1,300,000 Btuh.

"One boiler?" I said.

"Yes, the salesman claimed that it's crazy to have two boilers. He told me that there would be twice as much stuff to break down. He also said he was positive that this was the right size for the house because these two boilers have been here for all these years, and they've served the house well. After all this time, they *must* be right. That's what he said."

"But he's quoting 300,000 Btuhs over the total load of *both* boilers," I said.

"I know," the homeowner said. "The salesman mentioned that it's an old house, so it never hurts to have a little bit extra."

"Oh."

So I did what the salesman should have done, which was a heat-loss calculation. That's the only way we were ever going to find out the actual heat loss of that house. Anything less is just guessing.

And here's what we learned: The total load for the house on the coldest day of the year was 375,000 Btuh. That other boiler was a stand-by.

Why was it firing all the time?

My guess is that a service tech turned them both on one day and they stayed turned on.

From there, the abnormal just became the normal.

Happens all the time.

The homeowner decided to fire the oil company, which didn't surprise me. "They see my house as a vending machine," he said.

So that was that. I made him a sketch of how I'd like to see all the main gravity lines tied together into a primary loop, with the two boilers as a secondary loop, each loop would have its own circulator. It looked like this:

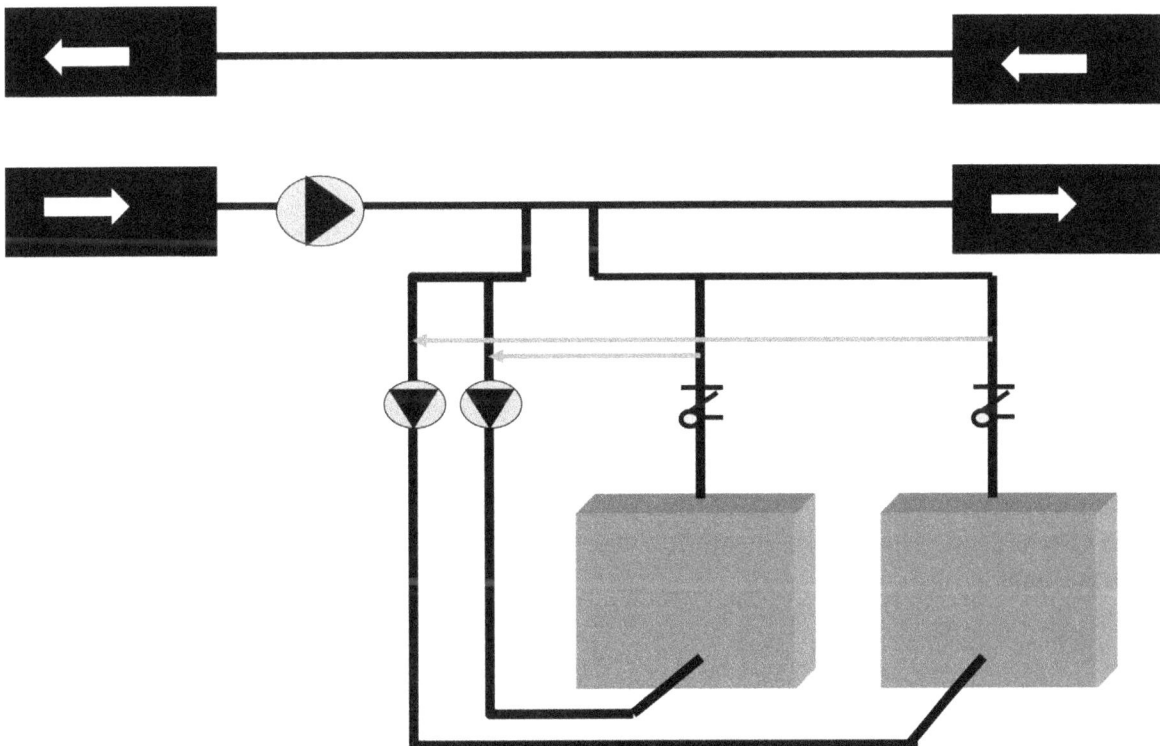

"Stop by a local supply house and ask them if they have in stock four, eight-inch-screwed by two-inch-sweat reducers," I said. "And don't take no for an answer. They'll try to jerk you around because you're just a homeowner. They've got them back in the warehouse somewhere. They just won't want to look for you."

Just kidding.

So with the supply-and-return gravity mains connected (with flanges and copper tubing, not reducers), I sketched how to use two boilers, with a *combined* load of 375,000 Btuh, on secondary circuits. I included a couple of bypass lines so the flue gases wouldn't condense. This story goes back to before we had condensing boilers; otherwise, I would have suggested those. As I told you earlier, condensing boilers just *love* those old, high-volume systems.

He had a local contractor install all of this and I followed up with him during the following winter. Most of the time, he ran just one of his new, relatively tiny, boilers, which is better than short-cycling a single 1,300,000 Btuh boiler all day long and on into the night.

Right?

So here's the big question: What do you think cost the oil company that account, and the contract for the new boilers?

I think it was laziness.

The salesman didn't want to take the time to do what a professional should always do when replacing a hot-water boiler. You have to do a proper heat-loss calculation. The salesman must have figured that if he went to the trouble of doing the calculations, and then didn't get the job, he would have wasted his time.

But you can see where that got him.

Thing is, I've been telling this story at seminars for more than 20 years now, and contractors still come to me during the break to argue. They say they can't afford to do heat-loss calculations on every job. It takes too much time. They'll go by the label on the old boiler. I was making them feel guilty. I have no idea what they're up against.

I'll mention that a proper heat-loss calculation will nearly always give them a smaller boiler than one you size by the Label Method, and that smaller boilers mean lower prices, and more closed sales. They'll argue with me about this too. It takes too long to do. They don't have the time.

"So how's business?" I'll ask.

"It sucks," they'll say.

So I think I'll just keep telling about Redflash.

You listening?

CHAPTER FIVE

Classic Pumping

As I was telling you, gravity systems were both the simplest and most-complicated systems ever to stroll through the history of hydronics. They were easy to understand but difficult to install. They were safe because they were open to the atmosphere, but that also meant that they were subject to lots of corrosion, and as they aged, they developed internal hydronic barnacles that slowed the water on its lazy climb to the radiators. So, in 1928, along came the American circulating pump, courtesy of Homer Thrush. (In Germany, a bit earlier, the pump company, Wilo, also came up with a circulator.)

Thrush circulator

Circulators circulated. They took the hot water from the boiler to the radiators and back again very quickly. They overcame the frictional resistance of the pipes and made heating more automatic. You could have a thermostat starting and stopping the circulator, and since oil was coming into its own at the time as a fine replacement for the drudgery of coal, life was getting easier for everyone who lived in a modern, hydronically heated home.

The first Thrush circulator was of the vertical variety. The motor sat above the pump body, and the pump had a packing gland that dripped water constantly. You had to adjust the packing gland to so many drips per minute, and you had to keep an eye on that drip rate as time went by. You also had to install the pump near a floor drain because it dripped all that water, and you had to replace that lost water with fresh water. It wasn't the easiest machine going, but it was better than what they had, that being just gravity.

The Dead Men installed those earlier circulators near the floor and always on the return side of the system because that's where the water was coolest. Cooler water was better for the packing gland, and these early circulators also had large flanges to accommodate the large pipe of a gravity system.

At the time, the Bell & Gossett Company was involved in the tankless water heater business. You'd hang one of those shell-and-tube heat exchangers on the side of your coal-fired boiler and let gravity take care of the rest. The B&G folks watched what Homer Thrush was doing with his circulator and decided to get into that business as well.

B&G had a smart guy named Ed Moore working for them at the time. Mr. Moore came up with the idea of calling the B&G circulator the "Booster" because that's exactly what it did. Add one of their boosters to an old gravity system and it would boost the heat to the building.

What's in a name? In this case, success. Smart marketing put B&G on the map.

B&G Booster

And they had one other advantage over that early Thrush circulator. The B&G Booster used a mechanical seal instead of a packing gland, so you didn't have to install this circulator near a floor drain. Mechanical seals do leak water, but it's such a tiny amount that it evaporates when it reaches the outer edge of the seal. It made things so much better.

They continued to install the circulators on the return side of the systems, pumping into the boilers because those early mechanical seals worked better with cooler water. The seal had a ring of carbon, which a spring on the circulators's shaft held tightly in place. The carbon ring spun against another ring, this one made of a material called remite, and that was the material that liked the cooler water.

Habit kicked in and installers decided that all circulators belonged on the return side of hydronic systems. Packaged boilers arrived and they came with the controls and the circulator installed by the folks at the factory. Naturally, they put the circulator on the return side of the boiler because that's where the circulator, with those first mechanical seals, would last the longest. And besides, that's where the installers expected them to be.

The customer is always right, right?

But then a funny thing happened. John F. Kennedy won the election, and shortly thereafter, declared in a big speech that the U.S. was going to put a man on the moon and bring him back safely by the end of the decade. NASA came into existence and they began figuring out how to get those astronauts back through the earth's atmosphere without setting them on fire.

Say hello to high-tech ceramics.

Ceramics can take very hot temperature and laugh at it. The people making pumps looked at these new ceramics as the perfect replacement for the remite in those mechanical seals. So, ever since the early-'60s, there has been no reason at all (other than habit) to install a circulator on the return side of any hydronic heating system.

But habit sure is a powerful thing, and there was nothing in the boiler manufacturers' self interest to move that circulator from the position it had taken in their packaged boilers. Think about it. The boiler manufacturer would have had to redo all their literature. They would have had to change their production line. They would have had to present something new to a group of customers who were reluctant to change. They would have been fighting a battle with absolutely nothing to gain, so they just left things the way they were and that was that.

Everyone was happy.

But meanwhile, something weird was happening in the world of commercial hydronics. Before World War II, most of the bigger buildings in the U.S. had steam heat. The post-war engineers were designing buildings during the boom times of the 1950s, and these people started to look lovingly at hot-water heat as a good alternative to steam heat.

At the same time, companies such as Bell & Gossett, Thrush, and Taco introduced the diverter tee, taking a lesson from the long-gone Oliver Schlemmer, of Cincinnati, Ohio and O.S.-fitting fame. The diverter tee made one-pipe hydronics possible, and this spelled the beginning of the end of the steam-heating era.

The challenge, however, was that diverter tees present a noticeable pressure drop to the water flowing in the main pipe. That Delta P is what drives the water out of the main pipe and into the radiators. And since diverter-tee systems are one-pipe systems, all the water has to flow through all of the diverter tees, so the pressure drop is cumulative. The more diverter tees you have, the bigger the pump gets.

Engineers started to specify these big, base-mounted pumps that sit on stout concrete pads. These engineers often oversized those pumps because they figured someone may want to add to the size of the building at some point. Hey, the '50s were boom times and they figured it was better to be oversized and ready for the growth that was sure to continue. Besides, electricity was cheap in those days, and who cared about energy conservation in the '50s? We were all going nuclear.

The engineers specified that the big pumps go on the return side of the system, of course. Habit, as we've agreed, is a powerful thing. Those pumps came on with a ferocious, Delta P wallop and pounded the water through all of those diverter tees.

Oh, and those pumps also started to suck air in from the automatic air vents, the valve stem packing, and even the mechanical seals. All that air caused many of those boilers to suffer oxygen corrosion, and as this happened, most engineers and installers scratched their heads.

They couldn't figure out what was causing it. But one man did.

Enter Gil Carlson

Gil Carlson passed away on April 28, 1994. He was 72 years old. At the time of his death he held seven U.S. patents and was recognized internationally as one of the foremost authorities on hydronic-heating systems. I had the chance to learn from him. He changed our industry, and he sure changed me.

He also moved the pumps from where they were.

The late Bob Dilg of Colorado's McNevin Company once told me a story about a time when he and Gil worked together at Bell & Gossett. This was during the early-'60s.

"One night after work, I found Gil standing in the parking lot. He looked bewildered so I asked if he was okay. He said, 'Yes, but someone has stolen my car!' I told him I would go back inside and call the police, but Gil said no, and asked if I could just take him home. Doris would have dinner ready, and she would be worried if he were late. He would call the police from there.

"As we drove up to his house, I noticed Gil's car in the driveway. Gill saw it as well, and without any surprise whatsoever, said, 'Well, I suppose Doris must have driven me to work today.'"

Just a day in the life of hydronic-heating's absent-minded professor.

I had a similar experience with Gil when he came to New York City to speak to a group of engineers. The day after his talk, I drove him from his hotel on the east side of Manhattan to LaGuardia airport, where he would catch his flight back to Chicago. I was dropping him off at the terminal when he turned to me and asked, "Did I have luggage?"

I had to drive back to Manhattan to get it.

I don't believe he had dementia. I think he just had other things on his mind, things that were more important than a suitcase.

We were sitting in an office in Manhattan one day when Gil told me a story about a problem job he had visited in that same city. This was in 1953 and hydronic heating was still in its infancy at the time. Gil was with Jack Hanley, who was Bell & Gossett's Eastern Field Representative back then. During their visit to this problem job, the two managed to come up with the concept of primary-secondary pumping. Gil went back to Chicago, sat at his desk, made notes, and primary-secondary pumping evolved from there.

The installer had used B&G's Monoflo® tees (their version of the diverter tee) on some perimeter radiation loops in this large office building. The problem, however, was that the pres-

sure drop through each Monoflo circuit was too high; water simply wouldn't move through the radiators. It's that Delta P situation again.

After a few calculations, Gil and Jack suggested that the contractor use small Booster pumps on each circuit, and run them, along with the main pump on any call for heat. The contractor took their advice and that solved the problem.

Today, primary-secondary pumping is a classic technique we use all the time, and one to which we hardly give a second thought. It's nice to stop and realize, though, that it all began with a problem job and some creative thinking.

Before it can become real, you first have to imagine it, and that's what Gil did all day long.

Gil graduated from Purdue University as an engineer and went to work at Bell & Gossett in 1946. He retired as their Director of Technical Services in 1988. He also served on the Industry Advisory Committee of Purdue's Herrick Laboratories for 32 years.

In 1953, Gil joined the American Society of Heating and Ventilating Engineers (now ASHRAE). Shortly thereafter, along with B&G's chief engineer, Harold Lockhart, Gil presented the paper, "Compression Tank Selection for Hot Water Heating Systems."

At the time, the Lockhart/Carlson paper represented breakthrough thinking in the science of hydronic heating. It greatly simplified the compression tank selection process and continues to this day to save countless hours of labor.

That first paper led to a second - the famous "Point of No Pressure Change" thesis, which proved that hydronic systems operate best when the circulator is on the supply side of the boiler, pumping away from the compression tank. When I'm done telling you about Gil, I'll do my best to try to explain the physics behind that thesis. It took me a while to get it through my thick skull, since Gil sometimes spoke to me in numbers, and I'm not so good with those. I'm better with English.

Gil believed strongly in the power of education, and he encouraged others to be all they could be. "I had no degree and Gil wanted me to progress," Bob Dilg told me. "He constantly encouraged me to further my education by taking night courses. When I told him that any degree would be years away, he just smiled and said, 'But you'll be smarter every day.'"

Gil could make a novice, such as I was, feel comfortable with even the more advanced systems because he had this way of coining a phrase. He would say things such as "A difference to be a difference has to make a difference." In other words, don't sweat the small stuff, and don't get too hung up on trying to solve a problem to too many decimal points. If it works, that's good.

He also used to declare in a profound way, "Whatever goes into a tee, must come out of that tee."

It took me years to realize the simple beauty of that statement, and how well it illustrates that Delta P represents the true Zen of hydronic heating. Look to the tee and you'll see the simplicity of hydronics.

"He was singularly, seemingly unimpressed with his own genius," recalled Bob Dilg. "One time I went into his office and he was holding a dirty piece of cardboard. It was probably from the back of his favorite yellow sketchpad. He had cut it in a circle and he had drawn some marks on it with a pen. I asked him what it was and he told me he used it for calculations. He never could find the book or charts he wanted.

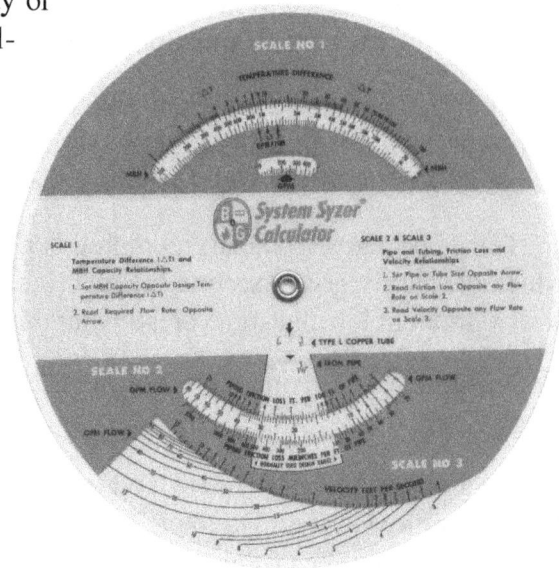

"In any event, what he had in his hand eventually became the Bell & Gossett System Syzer®. He just never realized that piece of cardboard he had cut and marked might be of use to anyone else!"

At one point, I was the caretaker of that round piece of cardboard that Gil had made. Gil had given it to his friend, Jim Hope, before dying, and Jim gave it to me. I gave it to Robert Bean, and it is now the Carlson-Holohan Industry Award of Excellence, (http://www.healthy-heating.com/Carlson_Holohan.htm).

Gil's Wheel, as we call it, has made its way through a number of worthy caretakers so far, and I think that would have pleased Gil. It certainly pleases me.

Once, while on a problem job in Philadelphia, Gil took another piece of cardboard, this time it was the cardboard tube from a roll of toilet paper, and used it (along with a slide rule) to solve a tough flow-balancing problem. That cardboard tube eventually became Bell & Gossett's Circuit Setter® valve. He had this wonderful way of seeing the magic and the potential in ordinary objects of everyday use.

Bell & Gossett Circuit Setter

Bob Dilg again: "He was so prolific that he would have overlapping ideas. He had so much to write down that he couldn't keep up with his own mind. He once tried a Dictaphone, but that only made things worse, since the secretary would type what he said. The ideas would still overlap and the typed copy rarely made any sense. It was part of one thought, mixed with another, and smeared over a third and a fourth. He went back to pen and pad.

"During the rough recession years of early ITT management (that conglomerate now owns Bell & Gossett), tough cost-cutting went on, and many people were let go or reassigned. These cuts would trouble Gil as he always knew there was plenty to do. His mind just churned out one idea after another.

"There was one event, however, that he never knew about. An ITT man saw Gil walking in the hallway, actually strolling in his fashion, with that faraway look in his eyes that signaled he was deep in thought. The ITT man raised the question among others as to who this person was, what he did, and whether he was necessary on the payroll since 'he seemed to lack urgency.'

"This, of course, was completely true. Gil never lived by his watch and sometimes had to be encouraged to go home. Thankfully, more knowledgeable heads prevailed and Gil's position was saved from the head-count reduction."

As time went by, Gil came up with improved procedures and charts for liquid viscosity pumping. He gave the hydronic engineering community a new way to balance flow by trimming pump impellers. He wrote papers on air-handling and antifreeze design, as well as on flow-to-heat-transfer relations. In all, he authored more than 100 published technical articles in *Heating/Piping/Air-Conditioning* magazine and the *ASHRAE Journal*. This is the stuff on which I was raised.

During the '70s, he developed new ways to design solar heating systems, cooling tower systems, and variable-volume, pumping systems. His fine mind ranged across a very broad field, but mostly he was a teacher. Over the years, he conducted hundreds of seminars, talks and symposiums throughout the world. ASHRAE honored Gil with their Fellow Award, their Distinguished Service Award, and in 1986, the prestigious Life Member Award. He was a member and chairman of numerous ASHRAE technical committees throughout his career, and he was also my teacher.

I will never forget him, and what follows is part of what he taught me.

I'll do my best.

A circulator is not a pump

A bicycle pump is a pump. So is an oil pump on an oil burner. When those machines start, you expect to get a pressure on the outlet side of the pump that's greater than the pressure on the inlet side of the pump.

A circulator is different because it's working within a closed, pressurized hydronic system. It doesn't have to lift the water to the top of the system because the water is already up there. The circulator doesn't lift anything; it circulates. It's very similar to the motor on a Ferris wheel.

The weight of the water going up balances the weight of the water coming down. The water is turning like a big wheel, and to get it going, all a circulator has to do is move the water that's inside of itself to the outside of itself. You can't compress water, so when you move one drop within a closed system, all the drops move in kind, and in the same direction. It's like moving a single link on a bicycle chain. All the links move at the same instant, right?

We call the circle at the center of an impeller the "eye." Delta P-wise, it's similar to the eye of a hurricane, but much friendlier. The impeller spins and creates centrifugal force. The water flows from the eye through the vanes of the impeller, creating a higher pressure at the tips of the vanes (which is also the outer edge of the impeller) and a lower pressure at the eye.

And there's that Delta sister again (P). A difference in pressure between two points will *always* cause flow. You know that instinctively. You watch the weather report on TV. They talk about a low-pressure area moving in. You know it's going to get windy because when there's low-pressure anywhere, air will rush over to fill in the hole. Or think about a tornado. It's a ridiculously violent version of an impeller.

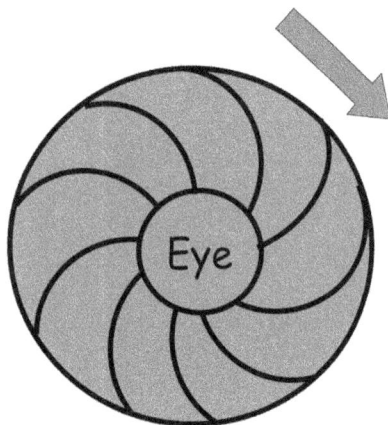

The impeller spins the water and directs it toward the outlet of the circulator, which is always a bit narrower than the inlet of the circulator. Not sure if you ever noticed that but it's true. Don't look at the size of the flange; look at the shape of the waterways entering and leaving the impeller. The quickly moving water experiences centrifugal force and suddenly has to race through that narrow exit, and it speeds up as it does so. That's what you're seeing if you have gauges on the circulator.

You can experience centrifugal force at any amusement park, or by taking that exit off the freeway too quickly. Ever notice all those black marks on the concrete barrier walls? They always make me think about hydronic heat, but that's just me.

I learned about centrifugal force when I was young and, as Mr. Hicks would say, *quite* stupid. I learned all about it with a circulator that was quite large.

I was working for that manufacturers' representative in New York City and we were trying to sell Bell & Gossett pumps to the wholesalers. As I mentioned, I was the Contractor Boy. It was my job to go out into the world of New York and help nervous contractors with things that had gone wrong. No one ever asked me to look at a job that was working just fine. Why would they? No, I spent my days looking at pure mechanical mayhem as a sort of hydronic policeman.

One day, a contractor called to tell me about this big, base-mounted pump that he had installed in an office building. "I put in what the engineer specified," he said. "But this one is real noisy. I think we may need a new pump."

"What sort of noise is the pump making?" I asked.

"It's shrieking like a jet taking off from Kennedy airport," he said.

"Sounds like velocity noise," I said.

"What's that?"

"It's what happens when a pump is too big," I explained. "The pump tries to shove too much water through the pipe and that's what makes the nose."

"But an engineer sized this," he said.

"They're human," I offered.

"I think we need a new pump," he said.

"Maybe not."

So I met him on the job and we put some gauges on this pump that was about the size of a sheepdog. I made a note of the difference in pressure between the pump's suction and discharge, and found that point on the pump's performance curve. This is easy to do because one psi of pressure is equal to 2.31 feet of head. Pump curves compare a circulator's flow rate to the system's resistance to flow, and they show that resistance as feet of head. So you just have to do a bit of calculating to find it. It's always Delta P times 2.31. I did that and sure enough, this circulator was way oversized, and it was proving that with its velocity noise.

"I still think we need a new pump," the contractor said.

"Maybe not," I answered. "Let's make a system curve here."

"What's that?" he asked, and I showed him how to do this. It looks like this.

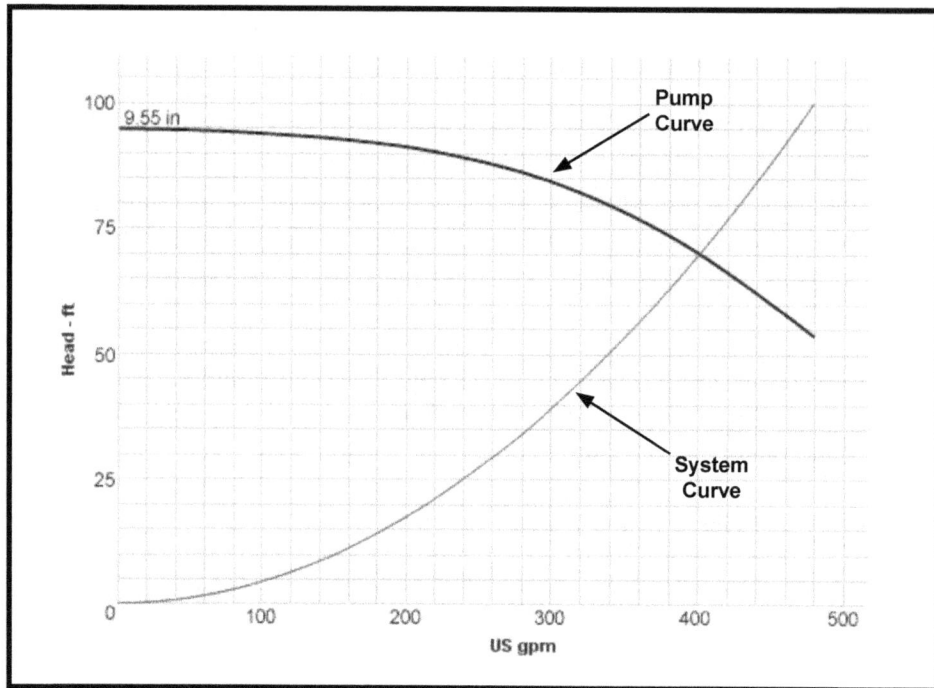

"This shows us where the circulator will operate if we trim the diameter of the impeller," I said. "We'll put it on a lathe and make it smaller. It's not my first choice as far as efficiency goes, but it will probably cure the noise problem."

"You sure I don't need a new pump?"

"Positive," I said. "In fact, this one is so oversized that we might just try running it without the impeller. Probably get enough action just from the shaft key."

"Huh."

"I'm joking," I said.

Well, with that bit of wit, he thought that I was just the brightest guy in the room, but he had never met Mr. Hicks, so he just started asking me all sorts of questions, and I was basking in the attention he was paying me. I felt like such a grown-up.

And I really did want to answer all his questions, but I first had to find a quieter place in which to do that, so this is what I did. Please follow these steps closely:

1. I shut off the big-as-a-sheepdog circulator.

2. I closed the isolation valves.

3. I removed the gauges.

4. I put the plugs back in the gauge tappings.

5. I restarted the big circulator.

Unfortunately, Step 5 was supposed to be: I *opened the isolation valves*.

I restarted the circulator should have been Step 6.

Oh well.

So there we were, this contractor and a much-younger me. We're standing just outside the boiler room. I'm dishing out my brilliance; he's listening. We did this for what felt like maybe seven minutes, but might have been a bit longer. It's hard to say. But it was right around then that we heard the explosion.

"What's that?!" he yelped, and it was just about then that I knew for certain what had happened because I had read about this phenomenon in a book. You (well, *me*, actually) are not supposed to run a pump with the isolation valves closed because energy in equals energy out, and if that spinning impeller can't move the water because the valves are closed, the energy of motion is going to turn into heat, and very quickly, leaving you with one very scary hydronic blender.

"What the heck was *that*?" the contractor shouted.

"Gee, I don't know," I said (which is exactly what *you* would have said). "Let's take a look."

We opened the boiler room door and what I saw made my sphincter do the mambo. Picture a rather large impeller spinning in space. The cast-iron volute that's supposed to encase the rather large impeller is lying on the floor in chunks. A piece of it is stuck in the wall.

I look at the contractor.

He looks at me

And then he says, "Didn't I *tell* you we needed a new pump!"

To which I instantly agreed.

And something tells me you would have done the same.

Later, I went back to my office with my tail between my legs – older, a bit wiser, and now very respectful of the power of bottled-up centrifugal force.

Walter Bosch worked for our company in those days and he was a very sharp mechanical engineer. He had come up through the Merchant Marine and he was delightfully profane. He knew just about everything there was to know about classic hydronics, and he loved hearing stories of mechanical mayhem because they offered him an opportunity to laugh explosively at kids like me, and to teach.

When I finally worked up the courage to tell Walter about what I had done, he let loose a huge guffaw and called me a dozen filthy words that only a sailor could know. He straightened me out but good, and then he wrote a memo to everyone in the company. I saved it because it made such a powerful impression on me then, and it still does.

Here is the gist of what Walter had to teach us that day in his memo:

"Recently, I was asked if a relatively large pump could run in a shut-off position for a full, eight-hour work shift. My response was, 'Hell no!' The person doing the asking quantified by explaining that there might be a couple of hundred feet of three-inch pipe connected to the pump. I said, 'Don't do it!'

(Walter was protecting me there by changing my story a bit.)

"I once attended a pump seminar sponsored by *Chemical Engineering* magazine. One of the speakers, the chief engineer for a major pump manufacturer, gave us a quick-and-dirty formula for calculating the temperature rise in the volute of a pump running with the valves shut off. Here it is:

$$\text{Temperature Rise (°F) per minute} = \frac{5.1 \times \text{BHP (at shut off)}}{\text{Volute volume in gallons} \times \text{Specific Gravity} \times \text{Specific Heat}}$$

"Let's assume the pump in question contains about seven gallons of water and has a 40-HP motor. Throw in 200 feet of three-inch pipe containing another 76 gallons of water:

$$\text{Temperature Rise} = \frac{5.1 \times \text{BHP}}{83 \text{ gallons} \times 1 \times 1} = 1.23 \text{ degrees/minute of operation}$$

"Now, 1.23 degrees per minute times 60 minutes times eight hours equals 590°F temperature rise. Neglecting the initial water temperature at saturation, we would approach 1,400 psig in the casing. Pumps don't have that much safety factor!"

Then he gave us some other examples of this phenomenon. The one that stuck in my mind was the simple Bell & Gossett Series 100, that little 1/12th-horsepower circulator that you'll find in so many old basements. Run a Series 100 with the valves closed and you'll get a temperature rise of just over 50 degrees per hour. Makes your mind swoon, doesn't it?

And consider a two-horsepower, Bell & Gossett Series 1535. That's the little close-coupled, end-suction, base-mounted pump you see on so many cooling towers all across America. Run that pump with the valves closed and you'll get a temperature rise of 1,590°F per *hour*.

Get it? I sure did.

Tough way to learn a lesson, though.

Be the water

So as we've agreed, a circulator in a closed, pressurized hydronic system is like the motor on a Ferris wheel. Its job is to turn the water, not lift it. It does this by moving a little bit of water from its inlet to its outlet. And since we can't compress water, when the circulator moves one drop within that closed system, every other drop also moves. It's like moving a single link on a bicycle chain. All the links move at the same instant, right?

Okay, now I have a few questions to ask you.

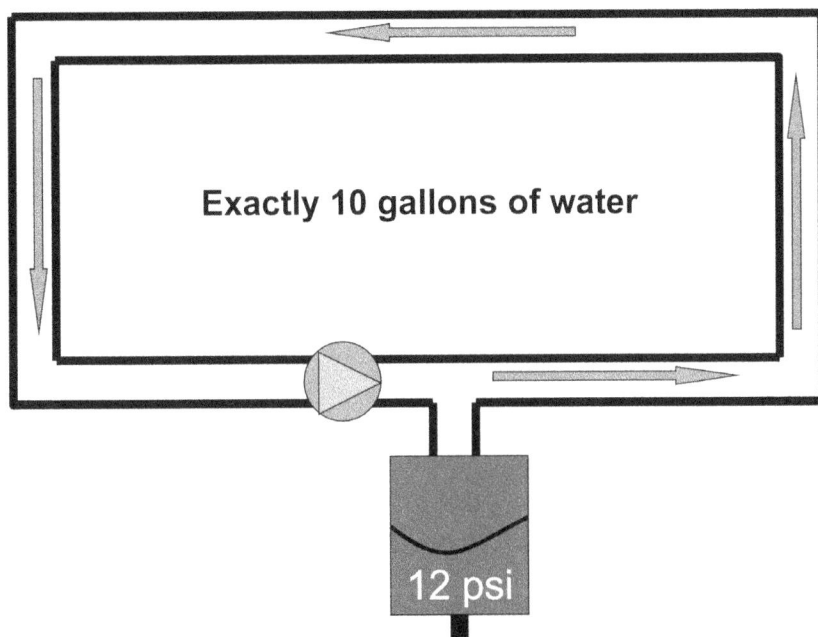

Exactly 10 gallons of water

12 psi

Here's a loop of piping with a circulator and a closed compression tank. I've removed all the other components we'd normally find in a classic hydronic system because I want you to focus only on the relationship between the circulator and the compression tank.

We filled this system with a hose through a water meter and then disconnected the hose and set it aside. We carefully measured the amount of water we needed to fill the piping and the compression tank until the pressure gauge registered 12 psi for the system. We purged all the free air. The water meter shows that this system now contains exactly 10 gallons of water. There's not a drop more and not a drop less.

Okay, now we're ready to move the water. We hit the switch and start the circulator. The impeller tosses out what's inside the circulator's volute and that causes all the water in the loop to move at the same instant because water is not compressible. It's going to move just like the chain on a bicycle.

Now here's the big question: When this water starts to move, will any of it enter the compression tank?

Look at the drawing and think about that for a moment. I'll wait.

It seems as if water *might* go in there, doesn't it? If you're thinking about that circulator as if it were a pump, it sure will seem that way. You're thinking that it must create an increase in pressure at its discharge, and that pressure will shove some of the water that's inside the pipe up against the compression tank's diaphragm.

You thinking that way?

Think a bit more.

And while you're thinking, keep in mind that there's only 10 gallons of water in the entire system. We can't add a drop more because we disconnected the hose that we used to fill this thing. Ten gallons is all we have. And if any of the water that's in the pipe leaves the pipe to enter the compression tank, what the heck is going to be in the place where the water used to be?

What's that? Did you say *air* will take the water's place?

Hmm, that's interesting. Where did you get the air? This is a closed system and we've already purged all the free air. Can you create air just by moving water?

Think it through.

If it were possible to make air appear by simply moving water, you and I could go out to the ocean and deep-sea dive with sump pumps. All we'd have to do is pump the ocean away from our noses and mouths and air would miraculously appear out of nowhere. Think of it. We could be super heroes!

Is that possible?

Think it through.

Can you have air appear out of nowhere just by moving water?

I don't think so either.

For the circulator to move even a single drop of water from that pipe into that tank, it would have to be able to do something that's impossible. It can't move water and create air, and it certainly can't move water and leave behind an empty space where the water used to be.

What's that?

You say the water will *stretch* to fill the empty space?

Really? Can you stretch water like taffy?

Oh, you're saying it will expand, as it does when we heat it?

But there's no boiler on this system. How are you going to heat the water to make it expand? All you're doing is pumping that water.

And consider this: Water is lazy. It always wants to follow the path of least resistance. When it leaves the circulator's discharge, all it wants to do is get back to the circulator's suction, and as quickly as possible because that's the point of lowest pressure. Keep in mind that high pressure goes to low pressure. Always.

Why would the water want to leave the loop and enter the tank? It defies physics.

Are you starting to see why it's impossible for the circulator to move water from the pipe into the compression tank? To do so, it would have to leave behind an empty space, a total vacuum, void of all matter. You'd have a slug of water, followed by pure outer space, and then another slug of water?

You think Mother Nature's gonna let you do that?

Even for a fraction of a second?

I don't think so either.

So I present to you an absolute fact of hydronics: When a circulator runs in a closed hydronic piping system, it cannot add a single molecule of water to the compression tank. And because it can't add water to the compression tank, it can't compress the air that's on the other side of the tank's diaphragm. And if it can't compress that air, it can't change the pressure that's inside the compression tank. That's because of Boyles Law.

Robert Boyle came up with his Law in 1662. He stated that if you have a fixed amount of air (such as you have inside a compression tank) and it's at a fixed temperature, which it is because we're not heating the water in this case, then the pressure and volume of the air will be inversely proportional, meaning that when one increases, the other decreases.

If the circulator can't add any water to the tank, it can't decrease the volume of the air on the other side of the tank's diaphragm by squeezing it. And if it can't squeeze the air, it can't increase the air's pressure.

Got it?

Great! Now let's look at this another way.

Exactly 10 gallons of water

12 psi

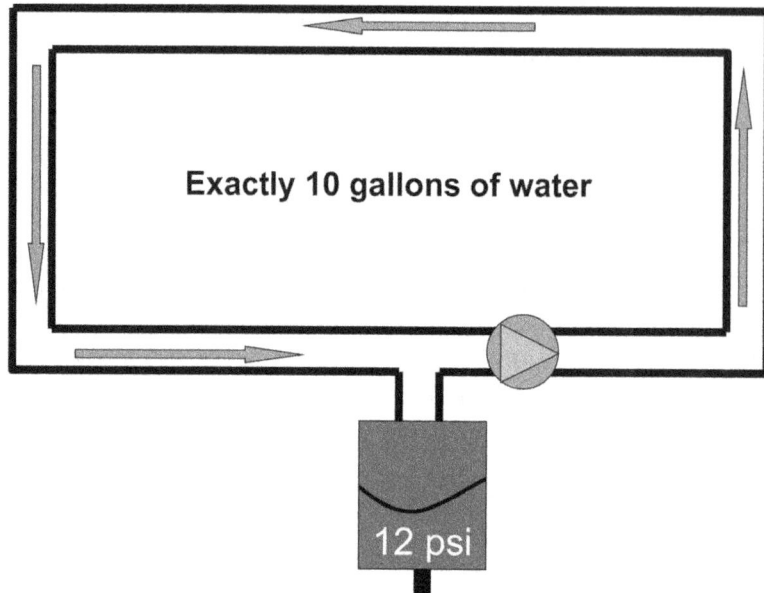

We repiped the circulator so now we're pumping *away* from the compression tank. The water is moving around the loop because the circulator is creating a difference in pressure between its suction and discharge sides. High pressure goes to low pressure, so the water flows through the circulator, and because it does, it flows everywhere else within the system. Remember, you can't move just one link on a bicycle chain.

So here's my next question: As the water zips by the connection to the compression tank, will any of the water that's in the tank leave the tank to join the flow that's moving through the pipe? In other words, will there be some sort of venturi effect that will suck the water out of the tank?

Mull it over.

Did you say yes?

Do you really think flow in the pipe can suck water from the tank?

Okay, I'll go along with that for now, but I have to ask how you did it. I mean the pipe is completely filled with water. So how did you add more water? You can't compress the water, can you?

So how did you get the water out of the tank and put it into the pipe if the pipe is already filled to the brim?

Did you say that the water that left the tank got replaced by water going into the tank from the other side of the pipe?

Hmm. Now you'd better think *that* through.

Why would water do that? It's lazy. All it wants to do is go from the circulator's discharge back to its suction. Why would it make its trip longer by taking a detour through the compression tank? It's not like there's anything interesting inside that tank for the water to look at. And the water doesn't need to bounce on the tank's diaphragm as if it were some sort of hydronic trampoline, right?

Think.

Can you see it? It's impossible for water to enter or leave that tank by way of the circulator. The circulator can't compress or decompress the air that's on the other side of the compression tank's diaphragm tank, so it can't affect the pressure inside the tank. That's why we call the tank, and the point where the tank connects to the piping loop **the point of no pressure change**. And that was Gil Carlson's big discovery.

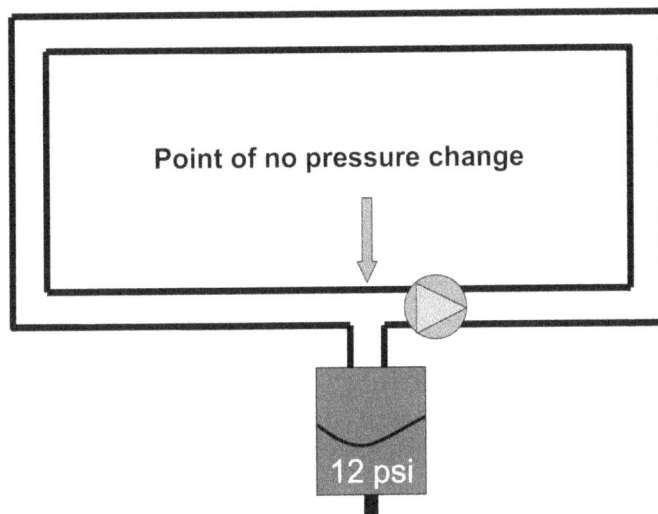

This spot in the system will always remain at whatever pressure you set the fill valve. If you're filling the system at 12-psi pressure, the pressure inside the tank, and in the piping that connects the tank to the system piping, will remain at 12-psi, whether the circulator is on or off. The only thing that can affect that pressure is the temperature of the water (because of Boyles Law). Heat the water and it will expand, causing some of it to enter the tank and compress the air. That will raise the trapped air's pressure.

The circulator can't do this.

Here's why this is important to you: If you pump away from the Point of No Pressure Change, you'll be adding the circulator's differential pressure to the system's static fill pressure. That will give you an overall increase in system pressure, and that will make it so much easier for you to move trapped air bubbles back to the air separator.

But if you pump toward the compression tank, you'll be subtracting the circulator's differential pressure from the system's static fill pressure. That leaves you with an overall decrease in system pressure, and the lower the pressure gets, the more likely it is that trapped air bubbles will ruin your day.

Here's an example that hits close to home

Consider a small, water-lubricated circulator, the kind you might have on your own boiler. Compared to those physically larger circulators from the old days, the modern circulator produces more pressure differential because it runs at a higher speed than the older version. Typically, the small circulator in your house can create a pressure difference of about six pounds per square inch. If you install that circulator pumping directly away from the compression tank, we'll probably see these pressures.

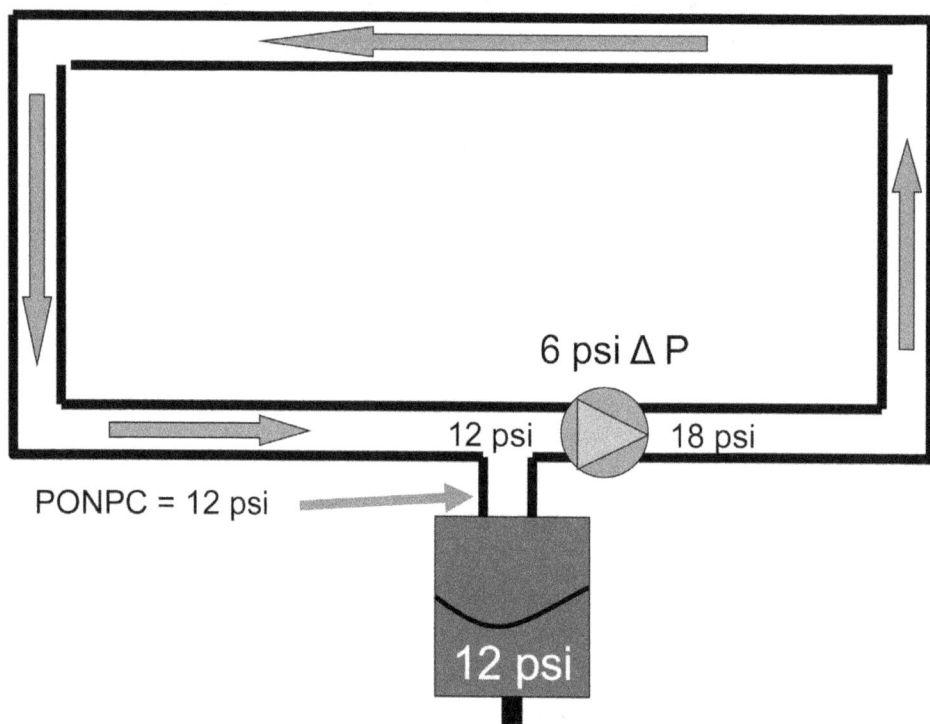

Notice how the circulator's full pressure appears at its discharge. This is because it's pumping away from the point of no pressure change. As soon as it starts, the circulator tosses out the water inside its volute and all the water in the system moves at the same instant because water is not compressible. Remember, it's like moving a bicycle chain.

The water flows from the boiler to the radiators, and when it gets back to the place where we have the compression tank, the pressure will be exactly what it was at that point when the circulator was off. That tank is the point of no pressure change. What's neat about this, though, is that we've added the full pressure differential that the circulator produces to the system. When you do this, Henry's Law kicks in to help you out. I'll tell you more about him later, but for now, just know that when you increase the pressure on the air bubbles that are upstairs in the radiators, you'll be busting them up, and making them so much easier to move. You'll probably never have to bleed another radiator. They'll bleed themselves.

And won't that be nice?

But back to this circulator and its pressure differential. Let's repipe it so that it's halfway around the system. We'll put it in a spot where the friction the flowing water experiences shows up on both sides of the circulator. Keep in mind that the tank is the point of no pressure change. To the circulator, it's like a hydronic bookmark, the place that will always be a constant. The circulator has to split its differential pressure to reflect the distance along which water must flow from the tank's location to the circulator's inlet, and then from the circulator's discharge back to the tank.

So it looks like this:

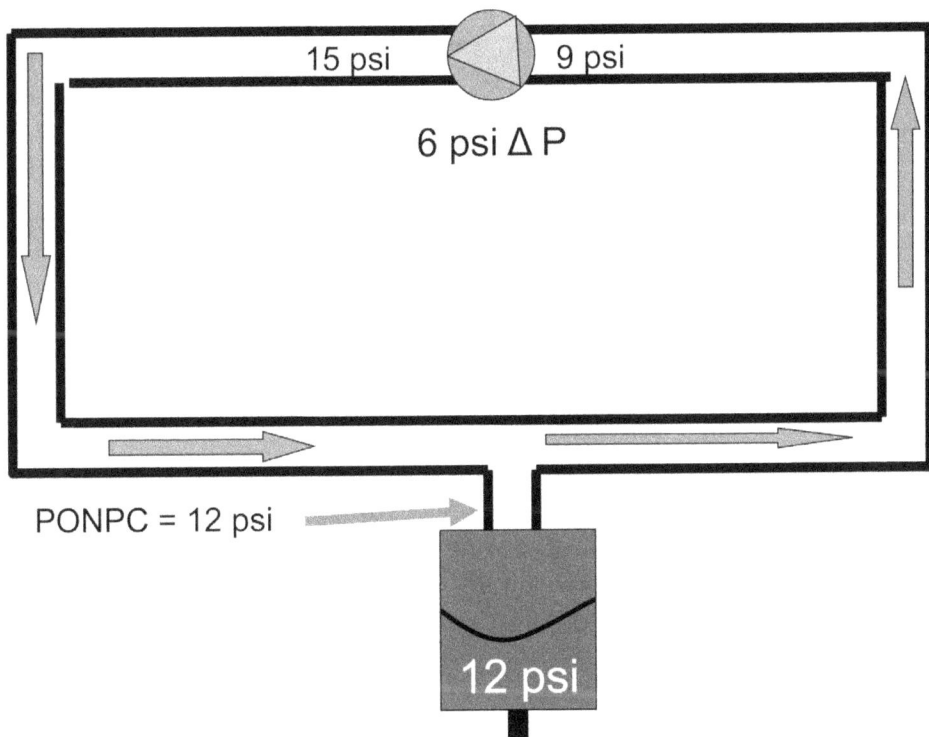

15 psi   9 psi

6 psi Δ P

PONPC = 12 psi

12 psi

The same amount of water is moving around the system because the system's resistance to flow hasn't changed. The pressure differential is the same overall, but since it's split between the circulator's suction and discharge, the pressure is now lower than it was on the circulator's suction side when we were pumping away from the tank. That can cause a problem, but not as big a problem as when we do this:

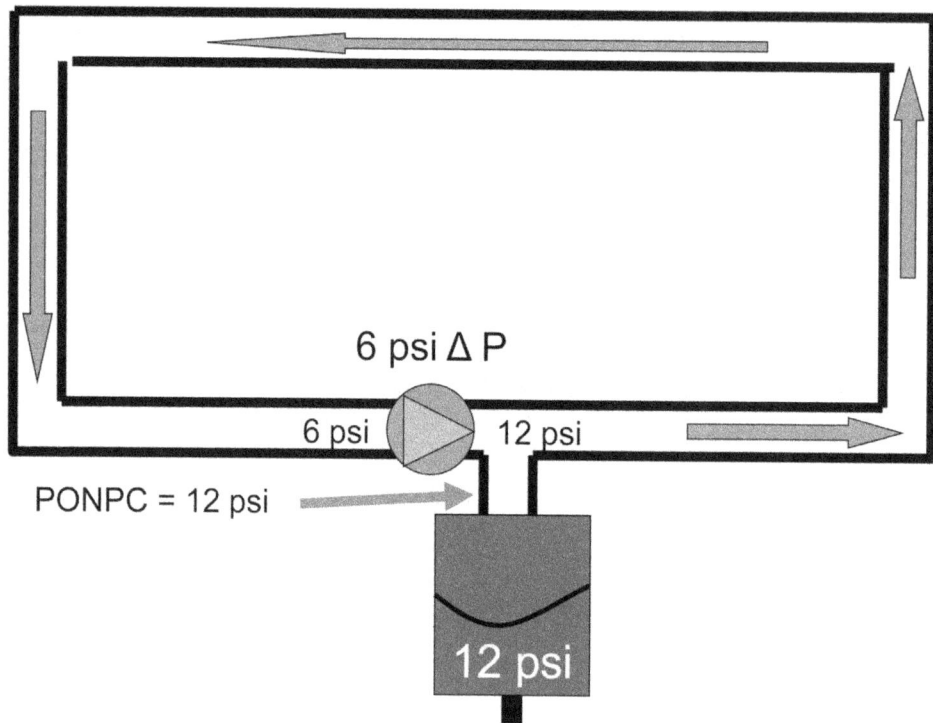

6 psi Δ P

6 psi

12 psi

PONPC = 12 psi

12 psi

We now have the circulator pumping right at the hydronic bookmark that is the point of no pressure change. The pressure at that point has no choice but to remain at 12 psi, even though the same amount of water is flowing through the system. The circulator doesn't care, though. It's still able to create its pressure differential, but look at how it does that. The entire Delta P is now appearing as a *drop* in system pressure at the circulator's inlet, and that's going to give you fits on the job.

Let's say you install a circulator on the return side of a boiler, pumping toward the compression tank (which is probably right there at the boiler's outlet). You're doing it this way out of habit. The person who taught you may have used those larger, slow-speed circulators. They didn't produce nearly as much pressure differential as what the circulators you're now using can make.

On this system, when the circulator starts, it's going to remove half the system pressure down there near the boiler. As soon as that happens, imagine what's going on inside those top-floor radiators. There are air bubbles in those radiators and Henry's Law just kicked in.

Want to know what that's like? Get a bottle of club soda and shake it up real good. Look at the bubbles that come out of the soda after you give it a good shake. That's carbon dioxide. The bubbles go to the top of the bottle and stay there. You might even notice how the pressure from those trapped bubbles is making the top of the cap push up a bit, but notice, too, how tiny the bubbles appear to be. That's because they're under pressure.

Okay, now hold the club soda right near your face (because I want you to remember this) and quickly undo the cap.

SURPRISE!

You just experienced Henry's Law, which, in plain English, states that a gas will dissolve in a liquid in direct proportion to the pressure and temperature applied. If you have a gas such as carbon dioxide dissolved in a liquid such as club soda, and you drop the pressure on the liquid, the gases will appear as bigger bubbles. The lower you drop the pressure, the bigger the bubbles will get, and big bubbles sure are tough to move through a classic hydronic system.

So you have to trudge upstairs and start bleeding those radiators.

And have you ever noticed how the radiator with the most air is usually the radiator that's behind the heaviest piece of furniture.

Life can be cruel that way.

Oh, and while you're moving furniture (a profession you did not choose), you're also stepping on the beige rugs, and the clean floors. And you're touching the white walls. You're doing this while the customer holds your final payment.

Now she doesn't want to give you that final payment because you've made a mess.

This is why it's always to your advantage to pump away from compression tanks. When you do that, you put Henry on your side. Instead of removing pressure from the bubbles inside the radiators, you add pressure to them, and that crushes them and makes them easier to sweep back to the air separator. You don't have to go upstairs and you stand a much better chance of getting that all-important final payment.

Life is always easier when you don't try to fight the laws of physics. It took me years to get that through my thick skull.

How about those larger systems?

Now I want you to think about a piping system you might find in a long, low warehouse, the sort of place that might have unit heaters hanging from the ceiling. Here's a simple drawing, minus a lot of the components, but with the stuff that's important to this conversation.

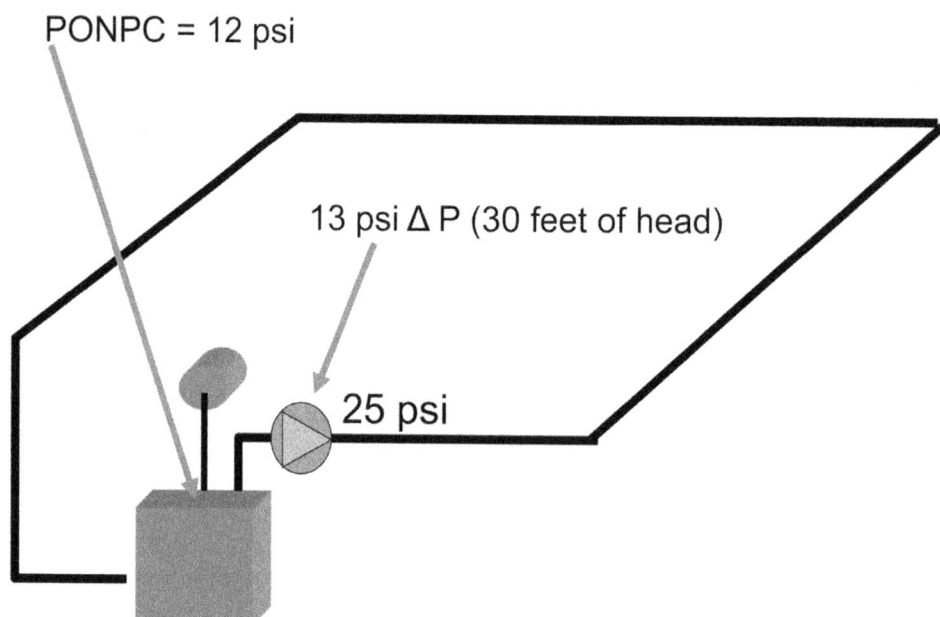

PONPC = 12 psi

13 psi Δ P (30 feet of head)

25 psi

This place isn't tall, so we don't have to lift water very far to fill the system. We measure the vertical distance from the fill valve to the top of the system, divide this by 2.3 (one pound of pressure will lift water that high), and then we add to that about three psi so that there's always pressure at the top of the system. This allows air to vent, and it also keeps the water from boiling at the top of the system, should you be working with high-temperature water. In this case, we're going to fill at 12 psi, which is just fine for a one-story building.

There's a long run of piping on this one, so our circulator winds up with a 30-foot head, which is equal to about 13 psi. Remember, this "head" has nothing to do with height; it has to do with the resistance to flow that the long run of piping presents to the circulator.

So we have a 12-psi fill and a circulator that can produce a pressure differential of 13-psi. That's going to make things interesting because the circulator's differential pressure is now greater than the system's static fill pressure. But don't be concerned because right now, we're pumping away from the compression tank (the point of no pressure change).

Notice how the circulator's differential pressure adds itself to the system's static fill pressure, giving us an overall positive pressure of 25 psi at the circulator's discharge. This is lovely because the higher pressure is going to squeeze the air bubbles and move them back to the air separator at the boiler.

Got it? Great!

Now let's introduce a bit of mayhem into this job. All we have to do is move the same circulator to the return side of the system and have it pump right at the compression tank. Watch what happens.

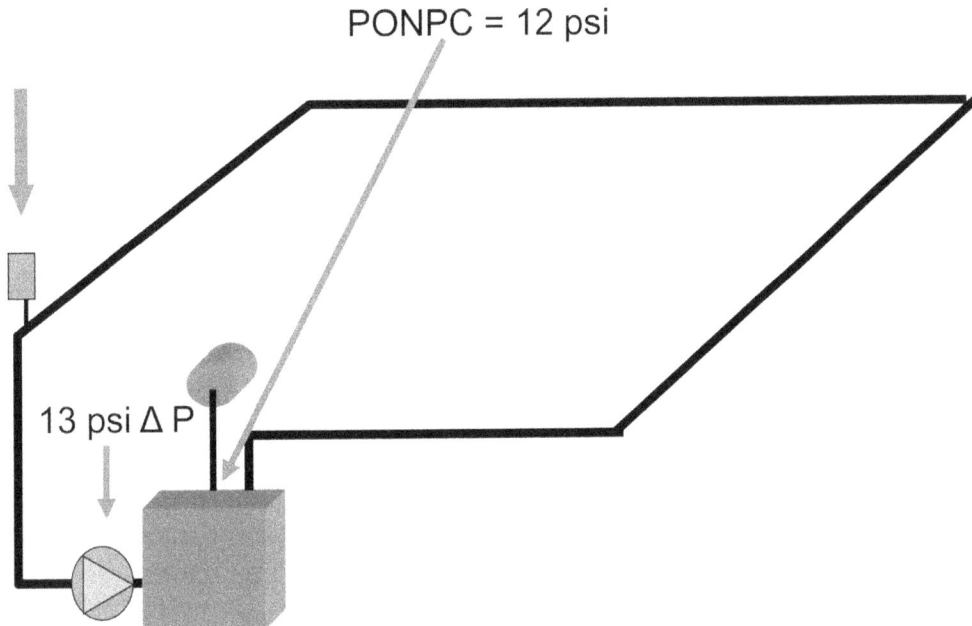

PONPC = 12 psi

13 psi Δ P

The point of no pressure change is still in the same spot as it's always been. That's the point where the compression tank connects to the piping system. The pressure drop across the system hasn't changed, so the circulator is still showing the same 13-psi differential across itself. But now the circulator is producing that differential pressure at a spot in the system where it can't show an increase in its discharge pressure because of the point of no pressure change. The circulator doesn't see this as a problem. It just shows all of its differential pressure as a drop in suction pressure, and this is where Mr. Mayhem enters.

The pressure at the circulator's suction is now below atmosphere pressure.

See that air vent at the top of the pipe, right there at the circulator's suction? Did you know that automatic air vents work in both directions? They can also be automatic air suckers, and that's what happens here. Air comes racing into the system as soon as the circulator starts. It flows through the circulator and enters the boiler, where it goes to work munching the ferrous metals. The air separator on the outlet side of the boiler spits out the air, so you might not notice there's a problem until the boiler dissolves from oxygen corrosion.

This happens a lot.

Hard Knocks

There is a school that most people in the trades attend at some point in their working lives. The entrance exam is when you take that first job. The school comes with no formal course catalog, but the tuition is often high, and you must pay.

It's the School of Hard Knocks.

Most people take the courses that school has to offer, pay the tuition bills, and learn. These people usually don't have to take the same course more than once, but there are those who never manage to graduate from this school. These are the ones who think they already know everything there is to know on their way in, which brings me to Smokey.

Smokey is currently on the other side of the lawn, but when he was with us, he was magnificent in his thick-headedness. Smokey was born with the entire world's knowledge already in his head and didn't feel he ever had to ask for advice. He would take on heating jobs with the total confidence that accompanies a thick skull. When things went south, he would continue to move in the wrong direction because that is the nature of a man who already knows everything. Physics, biology, mathematics and reason mean nothing to such men.

Smokey believed that cigarettes were harmless because his father had an uncle who smoked until he was nearly 90 years old, and if it was possible for that man to smoke for so many years without ill effect, then all the doctors in the world were wrong. For Smokey, the exception *was* the rule.

Smokey used one match in the morning to get himself lit off and this was only because he had to sleep for a while each night. If he could have kept last night's final cigarette burning through the wee hours, he would have used its still-glowing butt to get himself going at first light, but no one makes such a cigarette, so he used the one match and sucked smoke all day, as though it was a religious calling.

He asked for my help one day, but in a way that fit his character, meaning that he didn't actually *ask* for help (men who know everything require no help). He merely called to say that I really needed to see this horrible air separator, which was never able to vent all the air from this heating system that he had just installed because it wasn't as good as the ones they used to make.

"It vents whenever the pumps are running," he said. "It's never finished. I hate it, and you should hate it too. You should write about it in your books and tell everyone to stop buying this air separator because it never stops venting. It can't get the job done."

"Where is all this air coming from?" I asked.

"The air separator is making it," Smokey said, and there was not a trace of doubt in his voice. The man was born with his mind made up.

So we got together on the job, which had radiant tubing in a concrete slab. It was Smokey's first radiant job, but that was not a problem because Smokey already knew all there was to know about radiant floor heating. "I put the tubing in the concrete," he said. "The boiler heats the water, and the pump moves the water through the tubing. The building gets hot. I get paid. Case closed."

He was magnificent.

It was around 9:30 in the morning, and Smokey was smacking his second box of Marlboro's onto his left palm before tearing at the cellophane with his teeth. "These pumps suck, too," he said. "I want you to tell people about them. Tell them not to buy them."

"Why?" I asked.

"Look at how many you need to get the job done!" he said.

There were six of them, and Smokey had bolted them together, flange to flange, and on the return side of this cast-iron boiler.

"You get a special on these circulators?" I asked.

"That's how many you need to get anything moving through the tubing," he said. He lit another Marlboro and left the stub of the last one burning on the edge of the boiler.

"How much tubing do you have on this job?" I asked.

"I don't know," he said, "probably more than a thousand feet."

There were no manifolds. The boiler had a 1-1/4" supply, which Smokey had reduced to 3/8" after leaving the air separator (which never seemed to be able to finish the job). The single tube then entered the concrete floor like a deep-space voyager and eventually returned from the concrete on the other side of the boiler, where it connected to a commercial air vent that was the size of a can of beer. The first of the six circulators followed. From there, it was a hydronic daisy chain of pure pressure. It looked like the Six Flags of heating.

"You don't believe in radiant circuits, do you?" I said.

"Wadda ya mean?"

"A thousand feet of tubing, broken up into, say, five 200-foot circuits?"

"Too many chances for leaks," Smokey said.

"You buried more than a thousand feet of three-eighth-inch tubing as one long piece?"

Smokey lit another Marlboro, spit on the floor, and nodded. "Yep."

"That's why you need all these pumps," I said.

"I know," he said. "One's not strong enough. You should write about that."

"Oh, I will," I said. "And that's quite a boiler bypass."

Smokey has piped a half-inch copper line from the supply side of the boiler to the suction side of his circulator railroad. There was a ball valve in the line for balance. It was closed.

"Gotta protect the boiler from low temp," Smokey said.

"I'll bet the boiler comes up to temperature quickly."

"Like a rocket," Smokey said. "But you're here for that lousy air separator. Listen to that thing."

It was whooshing, all right. It was nearly as loud as the Budweiser-size air vent on the suction side of Smokey's circulator sculpture. He was staring at the air separator and shaking his head. "Garbage," he said, exhaling a cloud of smoke, which I watched swirl right into the big, automatic air vent.

"Have you ever noticed," I said, "how blowing and sucking sound about the same?"

"Wadda ya mean?" Smokey said.

"I mean how air, when moving quickly through a hole, makes about the same sound, whether it's going this way or that way. Blowing or sucking? Ever notice that?"

"Can't say as I have," Smokey said.

He blew some more smoke and the vent sucked it right in.

"Do you know about the point of no pressure change? How it's at the place where the compression tank is? And if you pump toward that point with a big enough circulator, or with, say, six circulators piped in series, stem to stern as you have them here, you're going to get quite a negative pressure on the suction side of all of that."

"Huh?"

"Blow some more smoke that way," I said, pointing to the big vent. He did. The vent sucked in the smoke nearly as quickly as Smokey was blowing it out.

"Wow!" Smokey said. "A system after me own heart!"

"Now watch over here," I said, pointing at the air separator. He did. "You won't see the smoke come out because this isn't a hookah."

"Huh?"

"You're sucking in air because you're pumping at the point of no pressure change with lots of pumps in series. The combined pump differential pressure is greater than the system's static fill pressure. That's why the vent is sucking. The air is probably eating this boiler. The air separator is trying its best to get rid of the air that the vent is sucking in," I said. "Sucking and blowing."

"I hate that thing," Smokey said, pointing at the air separator.

"Under the circumstances," I said, "I think you should be thanking it."

"I hate it. I need you to help me get even. You have to write about this," Smokey snarled.

"I will, when it's time," I said.

"When's that?"

"Probably soon," I said.

And now it's time.

On a smaller scale

Packaged boilers are popular, and these days, most of them come with a circulator that's in the carton, and not mounted on the boiler. I think that's a good thing because it gives the installer a choice. He may want to install the circulator on the supply side of the boiler, pumping away from the compression tank. Or, if he's like Smokey, he may do it the other way. Life is filled with choices.

The boiler manufacturers who sell the packaged boilers don't care where you install the circulators. They don't have to go upstairs to vent air from the radiators.

So let's say you have a packaged boiler that you're installing in a house and you decide to install the circulator on the return side because that's your habit. You also decide to pipe your automatic feed valve just above the circulator because it's easier for you to do it that way, and again, because of habit.

The compression tank, which is the point of no pressure change, is on the outlet side of the boiler, and maybe the job looks like the one in the photo on the right.

Okay, if you were the circulator, what would you do? You have to create a difference in pressure from one side of yourself to the other, but you don't have to necessarily increase the pressure at your discharge to do that. You could just as easily drop the pressure on your suction. It's six of one, half-dozen of the other as far as you're concerned.

You're pumping into the boiler and you have to flow through the boiler before you reach the point of no pressure change (the compression tank), so let's say you're able to raise your discharge pressure by about one pound to reflect the pressure drop through the boiler. But that's as high as you can go because you have to stop raising pressure once you hit the point of no pressure change. Think back to what the circulator did when it was halfway around the system from the tank. Remember how it split the difference, putting half the system pressure drop on its discharge side and the other half on its suction side? That's what's going on here.

You raise your discharge pressure by about 1 psi and then you get stuck. So you have to show the other 5-psi of Delta P as a drop in suction pressure. Take a look at this drawing.

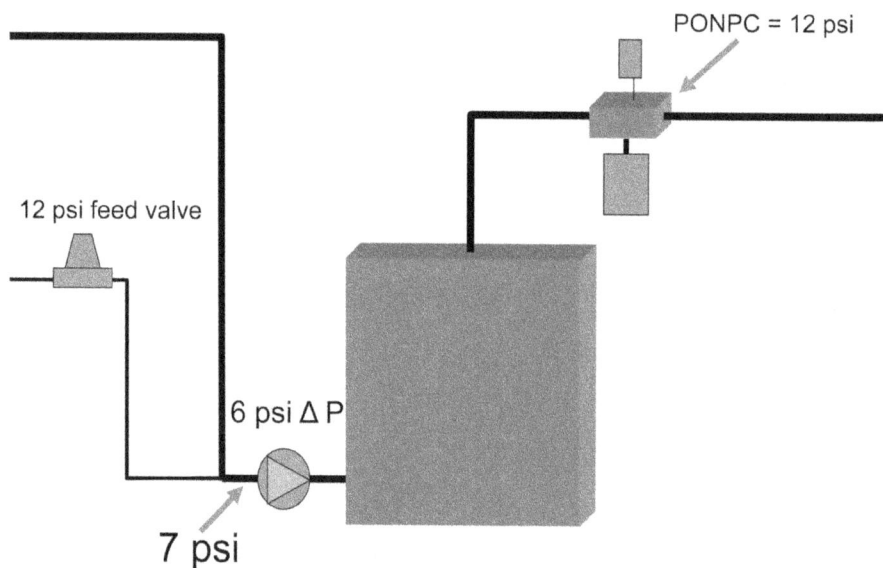

Check out where the 12-psi feed valve is. The circulator just came on and dropped the pressure by five psi right at that point. We're at seven psi now, so the feed valve is going to feed because you have it set to feed whenever the system pressure drops below 12 psi. It feeds because high pressure goes to low pressure. Always.

Now how often do you think this will happen?

Do you think it will happen every time the circulator starts?

It won't. It happens only once, the very first time the circulator starts. Think it through. On the first cycle, the pressure at the pump suction drops to seven psi. The feed valve feeds while the circulator is running, bringing the total system pressure up by about five psi. That extra water goes into the compression tank because that's the only place it can go. The rest of the system is already filled. The extra water in the tank squeezes the air on the other side of the tank's diaphragm, causing the overall system pressure to increase.

You've probably seen this happen. You fill a system at 12 psi, run it once, shut it off and now the gauge reads about 18 psi instead of 12 psi. You scratch your head and wonder how that happened. The next time the circulator starts, it again lowers its suction pressure, but this time, it lowers it to 12 psi because it's now starting at about 18 psi. The feed valve doesn't feed again because the pressure never dropped below 12 psi.

You're standing on the outside, watching this happen, and you figure it's no biggie because the relief valve on the boiler is good for 30 psi and you're just up to 18 psi or so. You shrug and scratch your head. Stuff happens, right?

Now the burner comes on and heats the water, causing it to expand. The "extra" water from expansion moves into the compression tank because that's the only place it can go (other than out the relief valve – which it will do in a moment).

You watch the pressure gauge climb and suddenly the relief valve starts to drip. You figure there's something wrong with that compression tank. Maybe it's too small. Maybe it doesn't have enough air in it. Maybe you should go get a larger one.

Okay, here's a hydronic fact of life for you: When you increase the system's fill pressure by 50 percent, the required compression tank size for that system *doubles*. So if you think the tank is now too small, you're right. It wasn't too small when the system fill pressure was at 12 psi, but it's too small now because the system fill pressure went up to about 18 psi.

You'll probably spend money on a larger tank, and all because of where the circulator is in relation to the fill valve and the tank. And you chose to pipe it that way.

Why would you do that?

A way to pipe that's better for you

I've been showing this drawing for a long time and a lot of installers have picked up on it, mainly because it costs less to pipe this way, and you don't have to bleed the radiators when you're done.

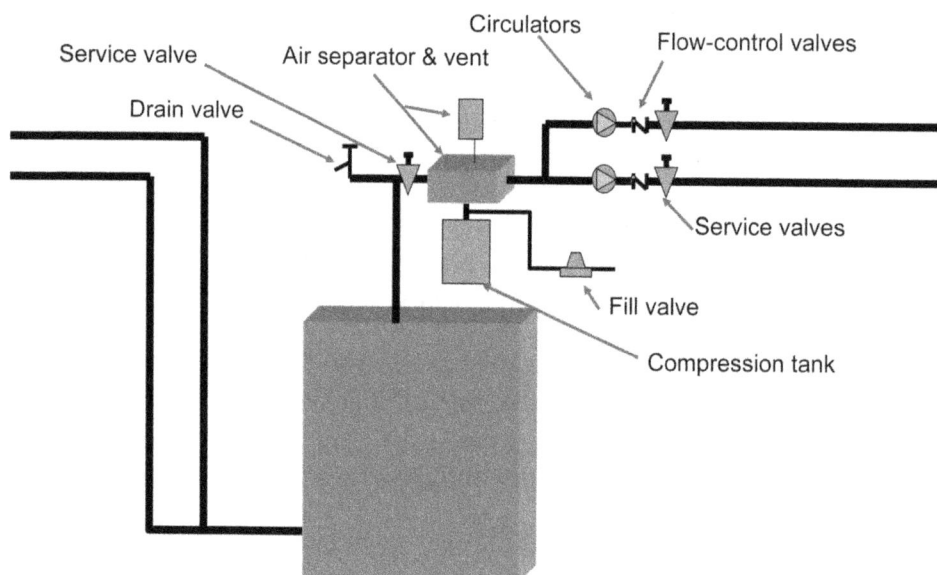

Leave the boiler with your main supply pipe and go into a tee. On one side of that tee, install a purge valve. That will be your purge valve for the entire system. After the tee, install a service valve. You'll use this to purge, and you'll also use it along with the service valves after your circulators when you have to service anything that's between those two points.

And you should check with some of the valve manufacturers because some of them have very cool, multipurpose valves that fit right into this picture.

After that first service valve, you head for your air separator. I'll tell you a lot about those in just a little while, but for now, install your compression tank at the air separator, and use this point for your fill valve because it's the point of no pressure change. It's the one spot in the system that won't trick the feed valve into feeding when it shouldn't.

Come out of your air separator and enter your circulator manifold. Chances are you're using the sort of flanges that let you isolate those circulators and that's good because we're going to start this up now. Those circulators may also have built-in, flow-control valves, and that will save you from having to add that additional component to your system.

Put a hose on the purge valve. Close the service valve after the purge valve and close the valves at all the circulators.

Open the feed valve and then open the valve at one of the circulators. Water will flow through the system, driving air ahead of itself. You'll purge the free air, and when it's gone, you'll close that circulator's valve and open the next one. Repeat this process for as many zones as you have.

When you're done, you open the valves and start the circulators. Since you're pumping away from the point of no pressure change, all of the circulator's differential pressure will increase the pressure on those air bubbles. Henry's Law will kick in; the bubbles will shrink and the flow will sweep them from the radiators and back to the air separator. You won't have to climb the stairs to bleed. You'll just have to go pick up your final payment.

Nice.

Some thoughts on feed valves

The feed valve's job is to, well, feed. As we've discussed, you figure out the height of the hydronic system from the point where the feed valve is to the highest hydronic component or pipe. You take that height in feet, divide it by 2.31 to convert it to pounds per square inch, and that tells you how much water pressure you need to fill the system to the top. And since you'll need some pressure at the top so you can vent air from that point, you'll add a couple of psi to what you just came up with.

If you need more than 12 psi to fill the system, make sure you check the air pressure in the diaphragm tank, and do this while there's no water inside the tank. The air pressure in the tank has to match the system's fill pressure while there is no pressure on the tank. And just because the tank manufacturer tells you there's a charge in that tank, check it anyway. You'll often find that what's printed on the box and what's actually in the tank are two different pressures.

But here's the thing: Since hydronic systems are closed to the atmosphere (or at least they're supposed to be), you shouldn't need that feed valve again unless you drain water from the system, which may make you wonder why you bought the thing in the first place, right? I mean some of the feed-valve manufacturers tell you to shut the valve once you're done filling the system. In fact, they do more than tell you; they plaster their installation-and-operation instructions with WARNINGS, which threaten you with the possibility of personal injury and/ or death should the valve remain open after you've filled the system.

That's enough to make Chuck Norris nervous.

But what if you close the feed valve and the air vents finally finish doing their job on those far-out circuits and the system pressure drops a bit? Now you have to go back to fill the system by hand, and that costs money. And since you'll be adding cold water to the system by doing this, and since cold water contains air, you'll probably have to go back more than once. Oh, the misery of it all.

You don't want to leave the fill valve open and defy the manufacturer's instructions (and all *that* implies); but you also don't want to have to keep going back to the job because that's just going to eat up your profit, aggravate your customer, and make you look and feel like an idiot.

So what do you do? Flip a coin?

Perhaps, but before you do that, I want you to understand why some of the manufacturers issue those warnings about leaving the feed valve open. This is a true story:

A house on a slab had baseboard heat and part of the copper loop dipped into the concrete to get by the front door. You've probably seen plenty of systems like that. On this job, the pipe that went into the concrete floor started leaking. That sent water down, and not up. It leaked for a good long time with nobody noticing because there was no vapor barrier under the slab. The water just drained away. The fuel bills went up, and so did the water bills, but the homeowners took it in stride because it happened gradually, as do so many things in life.

One day, the local water company decided to work on the main out there in the street. They went around the neighborhood, leaving notes in all the mailboxes, explaining that the water would be off for the day. The couple living in the house on the slab were at work so they really didn't pay much attention. They didn't see the work in the street as an inconvenience.

It was winter and the circulator was running while they were at work. There was no low-water cutoff on this boiler. The installer had depended on the open feed valve to keep the boiler and the system full, but with the city water shut off, the boiler ran dry. The thermostat kept calling for heat, and the burner kept running. Soon there was hardly any water in the boiler. The burner kept running, though, and the boiler got hotter than the hinges of hell.

At the end of the day, the guys working in the street were done and they opened the water main. Cold water spewed from the feed valve, hit the red-hot metal, flashed into steam, blew up the boiler, and took down a good part of the house. No one was home, which was a blessing, but this is one of the reasons why you'll see those warnings in the feed-valve instructions.

We learned from events such as this that all boilers need (and now *must* have) low-water cutoffs. It's insane not to use a low-water cutoff, and I know they sometimes lead to nuisance callbacks, but if a low-water cutoff trips a burner, that means it's time for a professional to take a look at that heating system.

I hope you agree with me on that.

We were kicking around this open-the-valve or close-the-valve issue on the Wall at HeatingHelp.com. Lots of people had plenty of points of view, and I read and considered each of them. And then Bob "Hot Rod" Rohr, who is one of the sharpest hydronics guys I've ever met, wrote this:

"If you have a rubber-tube radiant system, you either have a working fill valve or you may have to return every heating season to add a shot of fill water. For some reason, many of those systems tend to need a boost of pressure every year.

"I've tested some to 100-psi (tube only) for 24 hours and they held, but over the summer, they drop enough to prevent boilers with low pressure switches from operating.

"I suspect that for every disaster story of a fill valve left on, you could find one for a fill valve left off. Vacation homes are a classic case of freeze-up due to low-pressure lockouts that could have been prevented if the fill valve was allowed to do what it was designed and intended to do.

"It's your choice and there are plenty of arguments either way you go. It's a sad day when lawyers dictate hydronic installations. Which is the lesser of the two issues?

"Some auto-fill valves come with a knob to regulate flow. Once the system fills and purges, you can adjust the flow to provide some make-up without flowing four gallons per minute in the event of a break, and maybe that's a compromise?

"It can take days to purge all the air from a large radiant job. In a case like that, my advice would be to leave the fill on for a few days or a week. If you don't, you may be returning every day to top it off, only to meet an unhappy, no-heat customer.

"Perspective varies depending on whether you're manufacturing these, selling these, or on the receiving end of the unhappy customer calls. The customer just wants the reliable heat and domestic hot water that you promised, and billed them for.

"Maybe the tank-fill systems with alarm contacts tied into a phone dialer would be another answer. But whose phone number should we use?"

That's a good question, and this fill-valve situation remains an open end for the heating industry to consider. It's a classic conundrum, and I think we'll still be talking about it years from now.

CHAPTER SIX

Classic One-Pipe

Back in the day, before he became a business consultant and magazine writer, Al Levi, along with his dad and brothers, ran a contracting business on Long Island. We all became great friends. Al serviced an area of Long Island that was host to everything from old tenements to mansions. There were dozens of different types of heating systems, old and new, and it was a wonderful sandbox in which to play.

Al and I spent a lot of time roaming around basements, trying to figure out why this radiator wouldn't heat, or why that boiler was making noise. We were as tenacious as the seasons, and we never gave up until we had it figured out. I used to refer to Al as the Ace Troubleshooter, and he was that for sure. I still call him that today, but nowadays he troubleshoots other people's business problems instead of their mechanical problems.

One day, Al took me to look at a job where the baseboard radiator in the second-floor bathroom wouldn't heat. Here's Al with the radiator. Take note of that valve at the end of the radiator; it plays a cool part in this story.

Now, one of the nice things about being a contractor is that you have tools that can tear down ceilings, rip up floors and bust through walls, and most of the time, you're not working in your own house, so you're probably more inclined to use those tools while searching for a solution to a problem. Al figured the problem with this radiator started somewhere down in the basement because an Ace Trouble-

shooter always thinks like water, and always asks the key question: If I were water, which way would I go?

So here we are, back to the basement with the ceiling pulled down. Al's pointing at what we found.

You can see the horizontal main that bumps from corner to corner around the basement. Since this is a one-pipe system, that main is carrying both the supply- and the return water. It's like a highway. The hot water leaves the boiler, like traffic on the road. When it gets to a tee, it has to decide whether it should stay on the highway or head off through the bull of the tee and enter that other road, the one that leads to a radiator. Helping the water make that decision is our old friend, Delta P. A difference in pressure between two points always causes a fluid to flow.

So back to the photo. Notice those two tees that lead to the baseboard up there in the second-floor bathroom? One of them is a standard tee, but the second one (in the direction of flow) has a built-in restriction. Here's a cutaway view of what's inside that tee.

Okay, think like water. If you have a choice of going the inch or so from the first tee to the second tee, or up to the second floor, through the radiator, and then all the way back, what would you do?

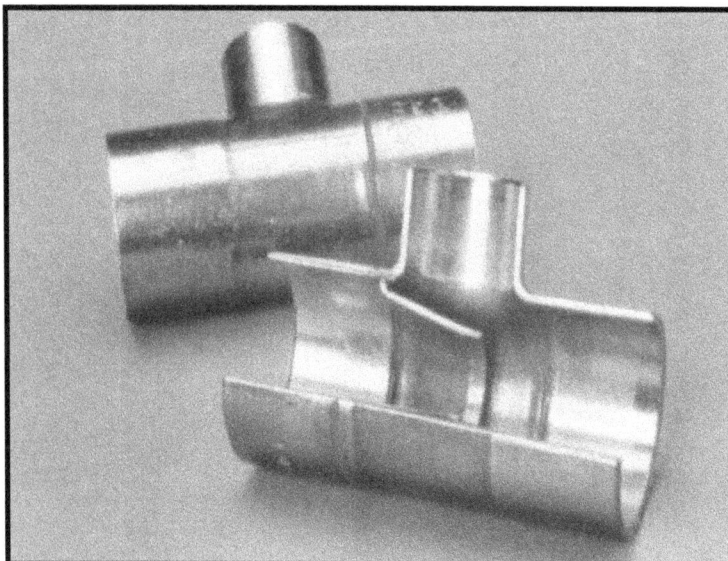

You'd probably stay on the main road, wouldn't you?

So would I. I'm water and I'm lazy. I'm always looking for the path of least resistance. I see a situation like this one and I'm thinking that it's easier for me to go straight, even though there's a resistance inside that second tee. I mean look at how far it is all the way up to the second floor and back. The heck with it; I'm staying in the basement.

So this becomes a no-heat service call, and it was making the guy who had the account before Al's company took over the job nuts. The woman who owned the house would call that guy because it was cold in her bathroom. The reason why it was cold was because there was no hot water flowing through her radiator. The guy didn't know this, though, and he did what most guys do when faced with a cold radiator. He bled it.

Oh, and you should know that the urge to bleed a radiator is greater than the sex drive in most heating contractors.

So he bled it and got no air while he was bleeding, and as I mentioned before, when you don't get any air you should stop bleeding because that ain't an air problem. It's a *balance* problem and the more you bleed, the crazier it's going to get because you're not actually bleeding; you're draining. Sooner or later, you'll drag hot water up from the boiler and the radiator will get hot. And you'll think you did something brilliant. This is because of the First Law of Hydronic Heating, which states:

When you do something stupid you will always receive a reward, which leads you to do things of even grander stupidity.

The guy would finish bleeding the "air" that wasn't there. He'd stand up, look at his hands and think, These *hands!* He'd bask in the praise of the homeowner and get in his truck. An hour or so later, she'd call to tell him the radiator was ice cold. He'd return and apply the same procedure because it had worked so well the first time.

Sound familiar?

After a while, the guy installed that purge valve on the radiator and left the woman with about four feet of garden hose. He showed her how to bleed the air (which wasn't there) from the radiator. All she had to do was stick the end of the hose in the tub and open the valve. Leave the valve open until the radiator got hot. Simple.

Here's where it gets good.

The woman explained to the guy that she liked to take baths, and she wondered whether she could fill the bathtub from the hose on the radiator. That would kill two birds with one stone, right? It would also kill the boiler through oxygen corrosion, but our guy was thinking only of how he could remove this woman from his life forever. So he said, "Sure you can! I mean it all comes from the same place, right?"

Well, not exactly. The boiler water comes from the wrong side of the domestic-hot-water coil, but we sometimes miss the finer points of hydronics as we search for ways out of town.

The boiler was failing when Al got involved and that was a nice sale for his company, and a mighty fine story for me, which I am delighted to share with you.

If you don't get any air, it *ain't* an air problem. Stop the bleeding.

The nature of diverter tees

But back to thoughts of traffic.

Imagine you're the water in the main. You're driving down the pipe when you notice an accident up ahead. There's a lane or two closed up there, so you hop on the service road.

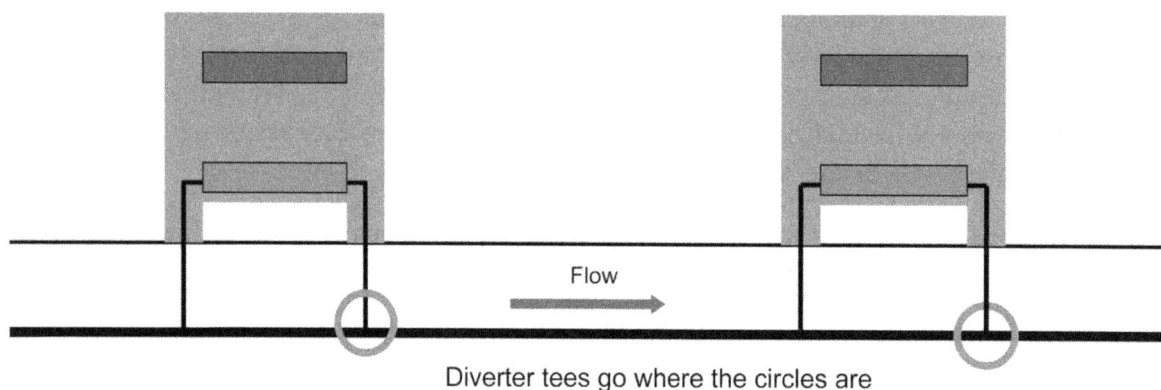

Diverter tees go where the circles are

But if that service road involves a 100-mile detour, you'll probably stay on the highway, right? Sure you would, and so would I.

That's why there was no heat in that second-floor bathroom. It looks just like an air problem but it's a flow-balance problem. If you don't get any air, it ain't an air problem.

Flow in a closed system is *always* about traffic. These tees don't "scoop" water. They just direct the traffic. If it's too congested on this main road, the water will get off and take that other road, but only if the detour isn't that long. The Delta P between the different ports of the tees decides where the traffic goes.

Can you see it in your mind's eye? Can you *feel* it?

Good.

Let's take a closer look at those tees. These are from two different manufacturers, but notice the ring on the run of the tees.

When you see these on a job, the rings should always be between the two pipes that run out to the radiator that the tees serve. Sometimes we use just one tee, and sometimes we need two tees. Here are the rules of thumb:

1. If the radiator is on the floor directly above the main, one tee should do the trick.

2. That tee should be on the return pipe (from the radiator), with the ring on the inboard side of the pipes that feed the radiator.

3. If the radiator is on the second floor, you'll need two tees, and the pipe leading to the radiator should be one size larger than what you would normally use. For instance, ½" for the first floor becomes ¾" if the radiator is on the second floor. This is to keep the pressure drop to and from the radiator to a minimum. Think like water.

4. The tees should ideally be the width of the radiator apart. When these tees were popular, so were freestanding, cast-iron radiators and convectors. These radiators and convectors were typically about three feet wide, and that's why you'll often see the tees placed at that distance apart on the main.

5. If you remove a freestanding radiator or convector and replace it with lots of linear feet of baseboard radiation, don't be surprised if the water decides to stay in the main. You just increased the resistance to flow by increasing the length of the detour. You made the side road longer.

Bell & Gossett

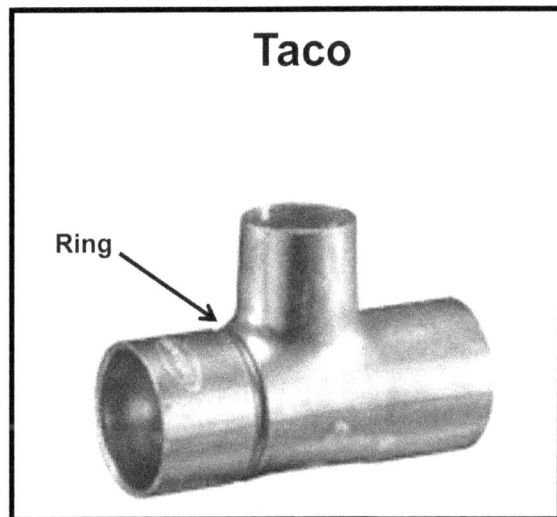

Taco

Ring

6. If the radiators are below the main, use a diverter tee on both the supply and the return, be certain that the rings on the tees are *between* the pipes that go to the radiator (meaning that the tees will face in opposite directions), and make sure the tees are as wide apart as the radiator is long.

7. If you remove a radiator and you're not going to replace it, connect the bulls of the two tees with a ½" pipe so that water has a place to go. Otherwise, you're leaving two major accidents on the main road, and closing the service road at the same time. That's going to slow the flow of traffic to the whole system.

One more thing: Staggering the tees (supply/return/supply/return) will increase the resistance to flow along the main and encourage more water to flow to the radiators. That's an old-timer's trick and it looks like this.

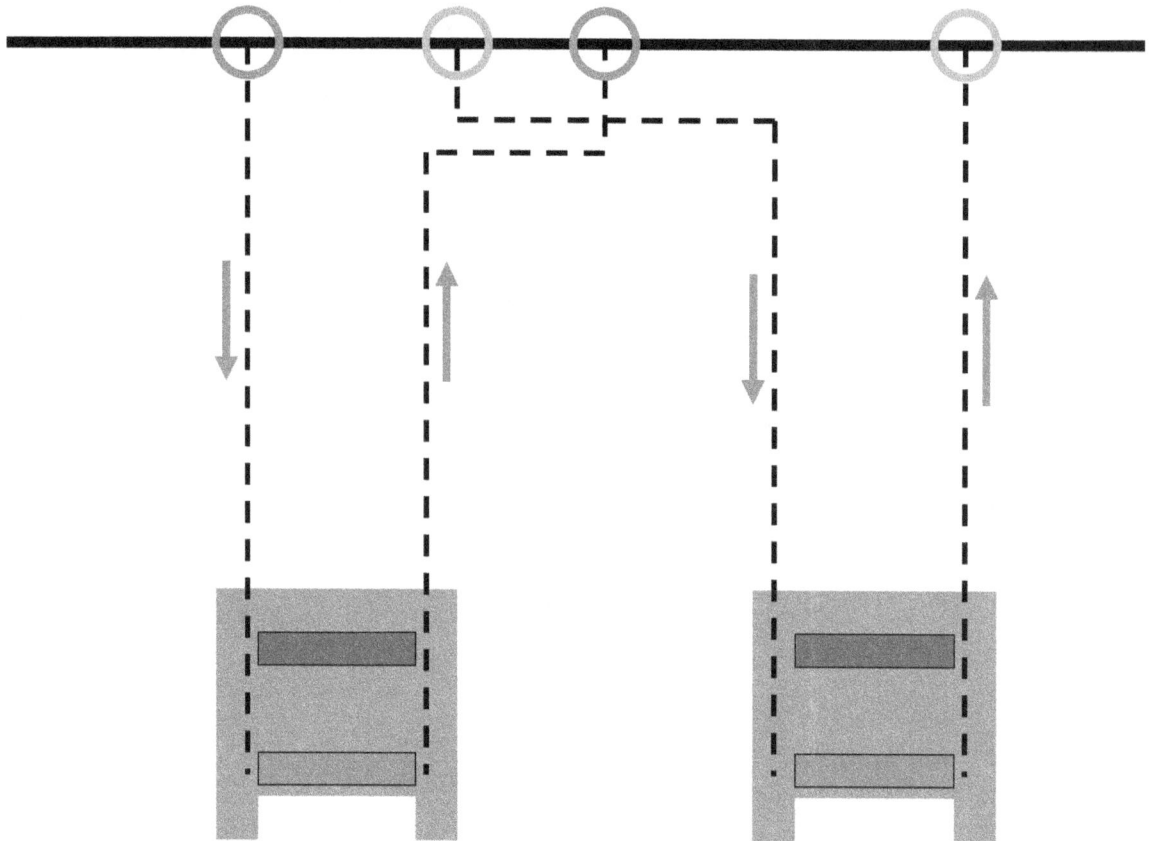

Thermostatic Radiator Valves and diverter tees

While I've got you thinking Delta P and how it can make or break a job with diverter tees, let's take a look at thermostatic radiator valves.

These nonelectric zone valves give you a way to zone one-pipe, diverter-tee systems and I like them a lot for that, but you have to be very careful how you use them. These valves, even when they're wide open, offer a resistance to the flow of water through the radiator, and that resistance might be enough to stop the water altogether.

The TRV has two parts. The part that attaches to the pipe is a normally open, spring-loaded valve. You'll attach to this the other part of the TRV, which is an operator that contains either a fluid or a wax that is very sensitive to changes in air temperature. As the temperature rises or falls, the fluid or the wax inside the operator will expand and contract, moving the spring-loaded valve open or closed. Control the flow and you'll control the heat. You can adjust a TRV to whatever temperature you'd like in the room it serves, typically between 50- and 90-degrees F.

What you need to watch out for, though, is the TRV's pressure drop. You can see this in the valve manufacturer's literature. They show it as C_v. That's an engineering term that always appears as a number. You'll see something like this: $C_v = 2.5$. That 2.5 is gallons per minute. Any number that appears after the = in the C_v equation will always be GPM, and what the equation is saying is that when, in this case, 2.5 GPM flows across that particular valve, there will be a corresponding drop in pressure from one side of the valve to the other of 1 PSI.

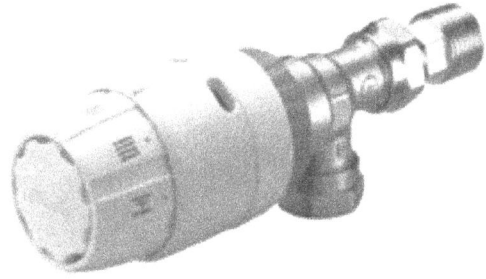

C_v *always* relates to a Delta P of 1 PSI. If you look at two valves, one where the $C_v = 2.5$, and the other where the $C_v = 3.0$, the latter valve will have less of a pressure drop.

Think it through. With the first valve, you get a 1-PSI drop in pressure with just 2.5 GPM flowing. The second valve can flow a full 3 GPM before the water suffers that same 1-PSI pressure drop. So if I were choosing between those two TRVs for my diverter-tee system, I'd probably choose the second valve because it has a higher C_v number, which means it offers less resistance to flow. I don't want the valve, when fully open and just sitting there, to present my flow with so much resistance that flow just stops because, where there is no flow, there is no heat.

Make sense?

A few other things to watch out for with TRVs

TRVs last a long time because they're so simple in design. They've been around for more than 80 years. In fact, the Empire State Building, which opened its doors in 1929, had them.

The TRV's bellows often has a thin, metal, remote sensor that sits out in the air like a mouse's tail. It feels the room's air temperature and operates the valve. Installers, who are sometimes handymen or building superintendents, will think nothing of mounting these crucial, room-air-temperature sensors, right on top of the radiator's element. They mean well; they're trying to protect the sensor, but their urge to help can lead to very memorable, building-wide, no-heat calls.

I've seen handymen and building superintendents jam the TRV sensors between the fins of the heating elements. How come? To hold them in place. These folks can be wildly creative in what they do. Some even use Krazy Glue.

They will also jam a TRV's sensor under a rug or a flap of linoleum. That keeps the sensor safe - and clean. Some handymen and superintendents mount the sensors on the radiator's supply pipe. That keeps them stable. They will also place the sensors precisely along that uncaulked crack between the floor and the wall. The low-bid building contractor left that crack as his legacy. This crack is the place where, in most older apartment buildings, Mr. Breezy slips in whenever the exhaust fans starts. No heat? Hmmm, I wonder why.

But enough about handymen and superintendents. Even professional heating contractors make mistakes when they snap those operators onto the valve bodies. You have to position them just so, and that's often tough when you're twisted like a yogi under the radiator and the sweat's running into your eyes. If you tilt the operator too much, the valve body's stem won't align with the part of the operator that's designed to receive it. They slip by each other, and from your position on the floor, you can't tell that the marriage wasn't consummated. You wind up with a valve body that's forever open. But everything looks fine from floor level.

The room goes to 90 degrees, and we get angry at the TRV manufacturers. But it's not their fault.

Make sure you're consummating that marriage.

And while you're troubleshooting, don't forget that engineers, too, are only human. I once looked at a problem job on Long Island where an engineer had specified TRVs for a perimeter loop in an office building. This guy called for 1-1/2-inch, commercial, fin-tube baseboard and a TRV in every office. The trouble was he didn't specify bypass lines from one office to the next. Only the woman in the first office was comfortable; everyone else shivered. Hard to believe, right? It happened.

And then there was that memorable day when the contractor installed about 700 TRVs in a hot-water-heated, Manhattan co-op. This was a two-pipe system. The engineer told the shareholders the TRVs would balance the temperature and improve their level of comfort significantly.

The contractor started the job, and the valves went to work. As the rooms got warm, the TRVs began to throttle, and as they did, they increased the resistance to flow. This shoved the big, base-mounted pump backward on its performance curve. Now as you reduce flow, you increase head pressure and that pump built plenty of pressure as it crawled backward up that curve. It didn't take long for all that pent-up power to shove the closed TRVs wide open again.

Seven hundred, brand-new TRVs and the building was overheating.

It hadn't occurred to the engineer to check to see if his new TRVs were going to be compatible with the existing pump. As time went by, we all came to appreciate flat, 1750-RPM pump curves and properly sized pumps. These days, we appreciate smart circulators and differential-pressure regulators for the same reason. As TRVs close, smart circulators slow down; differential-pressure regulators open up. Use either option and you won't get that big buildup in pressure at the TRVs, but I think the smart circulators are the better choice because they also save electricity.

Your call, though.

Square pipes?

I was in a hotel in Stockholm, Sweden when I first noticed the way the Europeans run their pipes. It was December, wicked cold outside, and there was a flat-panel radiator with a thermostatic radiator valve under my hotel room's window. The TRV was locked at about 65-degrees, so I used my Swiss Army knife to recalibrate it up to 90 degrees.

I'm quite the ugly American.

The pipes running to and from that radiator hugged the base of the walls and I thought this was really strange, being from New York and all. We're not used to seeing pipes inside the room. We seem to be ashamed of our pipes.

I met a Swedish engineer the next day and asked him about this brazen showing of their pipes. He told me that they install the pipes inside the room so that the pipes won't freeze, and also so that the heating professional can get at them, should there be a problem.

"In America, you put your pipes in the walls, don't you?" he said.

"Yes," I admitted. "We're ashamed of our pipes."

"You must have problems with freezing when it gets very cold," he said.

"We've actually built an entire industry around fixing frozen pipes," I said. "Many of our citizens depend on freeze-ups for their livelihood. It's an American tradition."

"That's ridiculous," he said.

But what does he know?

Years later, I found myself in a suburb of Frankfurt, Germany. A few friends and I were in the office of this big design/build construction firm. We were visiting the big ISH tradeshow in Frankfurt that year, and were on a side trip to get some insight into how the Europeans design their hydronic heating systems. The fellow who was lecturing us that day did so in German, but we had a translator with us, and he was a huge help.

This was the day I learned that the main difference between Europeans and Americans, when it comes to hydronics, is that they focus on pressure differential, while we focus on big flow rates.

They run little pipes from their boilers to manifolds, and then they connect their panel radiators to those manifolds. (See photo on the previous page.)

The radiators all have thermostatic radiator valves, and back then, they ran continuous circulation between the boiler and the manifolds. Nowadays, they use smart circulators with ECM motors that vary the motor's speed, based on the immediate flow needs, but these hadn't arrived yet when I was visiting. They were allowing the thermostatic radiator valves to modulate the flow between the manifolds and the panel radiators. They also used differential-pressure regulators between all their supply and return lines so that the TRVs could throttle the flow without causing the continuously circulating water to build pressure at the valves. Whatever didn't go through a TRV went through a differential-pressure regulator. That solved the problem of banging control valves and velocity noise, in the days before smart circulators. It was classic European piping.

During a break, I wandered over to one of the panel radiators that were under the window of this large conference room. (See photo above.)

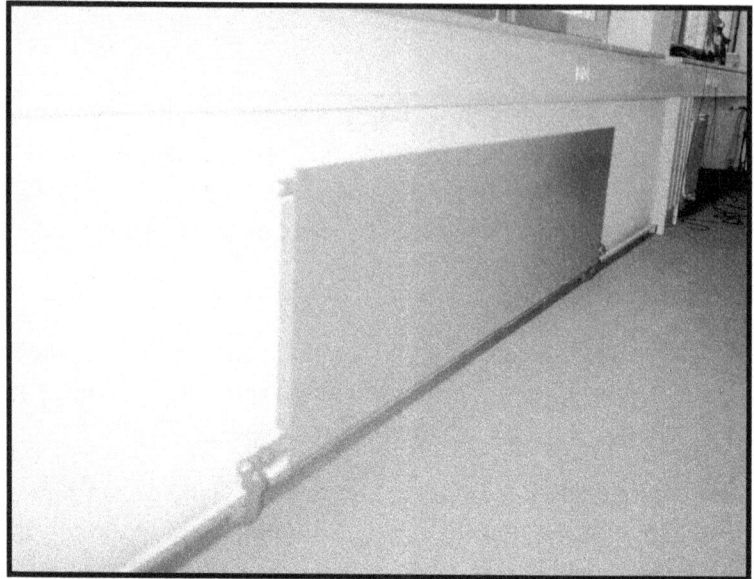

This radiator had a TRV, of course. The European Union law calls for them in rooms that are larger than eight square meters. The supply- and return pipes looked like they were going right through the wooden base molding and into the wall. Like the photo on the left.

I thought that was strange, especially since the building was old and the walls were made of stone, so I asked about it.

"Why do the pipes go into the wall?"

Our translator turned to our host and repeated my question. Our host looked confused. I pointed, gestured, and raised my voice.

"WHY DO THE PIPES GO INTO THE WALL?!"

He looked at me in just the way you would expect him to look at me, and then he explained, through our translator. "The pipes do not go *into* the walls. *That* is the pipe." And he pointed at what I had thought was base molding.

It wasn't molding.

It was a pipe.

"That's the pipe?" I asked. "That square thing?"

Translator, back to our host. Host nods.

"Why is the pipe square?" I asked.

Translator back to host, and then back to me. Are you ready for this?

"The pipe is *square* because the corner is *square*."

"Oh," I said.

"Square pipe fits in square corner," he further explained through the translator, as Mr. Hicks, up in heaven, shook his head sadly. *Stupid* kid.

"*Round* pipe, *square* corner, *not* so good," said our translator.

I just smiled like the silly idiot I was.

Turns out, this square pipe ran around the perimeter of the office building, feeding panel radiator after panel radiator, much as we would with a diverter-tee system here in America. One of my traveling companions looked at this and chuckled. "They must have some sort of diverter tee in there," he said. "Otherwise, water won't flow from the square pipe into the radiator."

I nodded. "I'll ask him about it." And so I did.

Now, our host didn't understand the concept of a diverter tee, so rather than raise my voice again, I grabbed a pen and paper and drew what I thought was a pretty good sketch of a one-pipe, diverter-tee system. It looked very much like that drawing I showed you before when we were talking about those things. I started to explain about hydronic traffic and all that, but it wasn't translating so well into German. Our host just looked at what I was drawing and then he mumbled something in German and scowled.

"I don't think he likes you," my buddy said. I stepped back.

Turns out, I was showing our very-German, and extremely energy-conscious host an American system that is designed to waste lots of electricity by having cumulative fixed resistance built right into it from the get-go. Diverter-tee systems lead to high electricity bills because the circulators have to fight their way through all that resistance. But in America, we have long believed that bigger is better, electricity is a cheap bargain, and that we're right.

So there.

This wasn't going over well with the Germans.

Oh, and this was also the day when I learned that the pressure drop through a panel radiator is so slight that the water favors it over staying inside the square pipe. When the water is in a panel radiator it thinks it's in the North Sea.

Who knew?

So there we were. I was smiling. The German was scowling, and my buddy was drinking coffee. Under his breath, he says, "You know what, Dan? This job can't possibly work because the water from the first panel radiator is cooling the supply to the next, and so on. Just look at it. They can't run supply water hotter than about 155 degrees in this country. It's the law. So how the heck are they going to heat these offices especially the ones at the end of the loop?"

"I don't know," I whispered.

"*Ask* him," he whispered back.

"*You* ask him," I whispered.

"Why would I do that?" my friend hissed. "He already knows *you're* stupid. Why get me involved? Just ask."

I mulled it over.

That's one of the biggest challenges with one-pipe hydronics. The supply- and return water share the same pipe, and the water just gets cooler as it travels from radiator to radiator. This is especially critical when you have diverter tees involved. The return lines from those tees keep dumping relatively cool water into the hot (and getting cooler) supply water that's heading to the next radiator.

This German system had very long loops running around the perimeter of the entire office area, and I was thinking about Delta T, so I took a deep breath and tossed it out there. Then I waited to hear what my new German friend, who thought I had the intelligence of a turnip, had to say.

But he didn't say a thing. He just smiled. Then he motioned for us to follow him, which we did. We wound up in the basement and down in that basement was the most wondrous hydronic bouquet I have ever seen. It wasn't complicated; it was brilliant. It was a circulator, a four-way valve, a switching relay, and a timer, and here's how they worked together:

The circulator comes on and runs hot water clockwise through the four-way valve and the loop of square pipe, feeding all those panel radiators along the way. The end of that loop is getting cool, and those final radiators are not so hot. So after a half-hour of flowing clockwise, the timer shuts off the circulator and the relay reverses the position of the four-way valve. Then the circulator restarts, but now it's moving the water in the *opposite* direction around the big loop. The supply water goes this way and then that way, and then this way, and everyone upstairs stays cozy.

And isn't that a wonderfully simple solution to what might have been a serious Delta T problem?

Live, learn, and hopefully, get smarter every day.

Oh, and try not to disappoint Mr. Hicks.

Loop systems

In 1950, they built the house we're living in now and gave it a copper-in-concrete radiant heating system. In 1970, the previous owner noticed wet spots on his kitchen floor and knew that the family budget was in trouble. There's not much you can do when the copper in the concrete rots and springs more leaks than a litter of puppies.

So the guy called a heating contractor, who showed up and pronounced the radiant floor heating system dead. The contractor gave the guy a price to convert the system to copper, fin-tube baseboard. He didn't do a heat-loss calculation on the house. He just sized the baseboard by installing it everywhere there happened to be a wall. For instance, a 10-foot by 15-foot room would get 25 feet of live baseboard element. He skipped only the front and back doors because they don't make baseboard with hinges.

Most contractors on Long Island do this because it's convenient. There's no basement in these homes, and the contractor needs a place to run the loop. In most cases, there's no need to run live element everywhere, but a fear often sets in. What if I don't give the customer enough element? What if they complain next winter and I have to go back? Better that I fill it up with element now. After all, they're paying for it, right?

The upstairs of our house, not coincidentally, has just as much baseboard as downstairs. If you ever need any, give me a call.

Here's the best part. Many of the contractors I've met will size a boiler based on the amount of baseboard that's in the house. They think this is as good as (and much quicker) than doing a proper heat-loss calculation. It's not, and to prove it, here's a conversation I had with a contractor friend not too long ago:

Me: "How do you size boilers?"

Him: "Hot-water boilers?"

Me: "Yes."

Him: "I go around and measure the radiators."

Me: "Why?"

Him: "You gotta support the radiation."

Me: "But you're probably going to oversize the boiler if you measure the radiation on a hot-water job."

Him: (laughs): "I've been doing it that way for years."

I invited him to my house and told him I was thinking about a new boiler. I asked him what size I needed. He took out a tape measure, and calculated that our house has 200 linear feet of 3/4" baseboard. He then told me that each foot of baseboard puts out about 580 BTUH when the water is 180 degrees. He multiplied that number by the number of linear feet and came up with a boiler size of 116,000 Btuh. To this he added a bunch more Btus because we have an indirect domestic-hot-water tank and four daughters.

Our house has four zones and when I did a heat-loss calculation, I came up with 40,000 Btuh, and that's for a 15-degree day with the wind blowing.

My buddy told me it would never work.

You gotta support the radiation.

Whenever my friend loses a job he blames it on customers being too cheap.

Even though he quotes boilers that are about four times larger than they need to be.

But I digress.

Baseboard loop systems run from room to room and the water keeps getting cooler as it goes along. That's why we have to put limits on the lengths of those loops. Here's a good rule of thumb:

If the baseboard is ½", keep the total linear footage of element under 25 feet.

If the baseboard is ¾", keep the total linear footage of element under 70 feet.

If the baseboard is 1", keep the total linear footage of element under 104 feet.

That doesn't include the bare pipe to and from the element. Bare pipe gives off heat, but not nearly as much as pipe that has fins. The limits above are real, and they come from the baseboard manufacturers. If you push those limits, you'll run out of heat. Not literally, of course; the water will still be warm, but it probably won't be warm enough to heat the rooms on those really cold days, especially if those rooms have doors that close (such as bedrooms and offices do). The last rooms will be cooler than the first rooms, and it will seem to be an air problem.

It ain't.

You can fiddle with the dampers on the baseboard if the system has too much element. Closing the damper limits the amount of air that can come in contact with the hot fins. That slows the convection of the air and lessens the amount of heat going into the room. But then you have to wonder, once again, why there's so much element to begin with. I mean why use it if you're only going to close the dampers?

Oh, and just so you know, if you remove the front cover of the baseboard, you won't get more heat; you'll get less heat. That cover promotes the convection of air over the hot element. Without it, you get less convection, and less heat.

If you're working with a wall-to-wall, baseboard-loop system such as I have, you can take advantage of it by lowering the boiler temperature. Here's what I mean. Let's say you have a room with a heat loss of 5,800 Btuh. If the installer put in 10 feet of 3/4" baseboard and supplied it with 180-degree water, he'd match the heat loss on that cold day because 3/4" baseboard puts out 580 Btuh when there's 180-degree water flowing through it.

But suppose the installer put in 15 feet of 3/4" baseboard instead of ten. With 180-degree water flowing through it, 15 feet of 3/4" baseboard will put about 8,700 Btuh into the room. That's way too much - even on the coldest day of the year. People will be uncomfortable.

But since this extra baseboard is already there, you can run water at 150 degrees. Under those conditions, each foot of 3/4" baseboard will put out 380 Btuh. And since you have 15 linear feet, you'll be emitting 5,700 BTUH into the room. That's just about right for the coldest day of the year. It costs less to operate at 150 degrees than it does to operate at 180, right? That's certainly in the customer's self-interest, and if you're a contractor, it gives you a great talking point.

And just to be fair, I should mention that not every installer oversizes baseboard. Here's an example of that, posted on the Wall at HeatingHelp.com. And doesn't this just make your heart flutter?

Oh, and condensing boilers love over-radiated houses and that low-temperature return water. Check out some of the modern controls that will automatically lower the water temperature in an over-radiated house. It's like putting cruise control on that classic loop system. These controls also minimize the amount of expansion and contraction noise you can get from copper baseboard when it heats and cools.

One-pipe, diverter-tee systems and loop systems have their limits, and that's because of our friend, Delta T. The water temperature just keeps dropping as the radiators give their heat to the air. And this is why the Dead Men came up with the idea of two-pipe hydronic systems.

Let's take a look at those classic systems.

CHAPTER SEVEN

Classic Two-Pipe

You can send hot water through an insulated pipe a long way without having to worry about it being too cool to get the job done once it reaches the radiators. This is especially true if cooler return water from radiators isn't blending into the hot supply water along the way, as it does in diverter-tee systems. Every radiator gets essentially the same temperature water in a two-pipe system, and that's its greatest advantage over one-pipe.

Balance also comes into play here, and that's why we have two basic types of classic two-pipe systems. The first is direct-return, and the second is reverse-return. They work like this:

Direct-return

Direct-return looks like a ladder. (See the drawing on the next page.) There's a supply main for the hot boiler water and a return main for the cooler return water coming out of the radiators and heading back to the boiler. The radiators are like the rungs on a ladder. Notice how the return water doesn't blend together until after it's done its job in the radiators.

The challenge with two-pipe, direct-return, however, is that water wants to follow the path of least resistance, which will be that first circuit, the one that's closest to the circulator. Keep in mind that water is lazy and it's always looking for the shortest (lowest-pressure-drop) path from the circulator's discharge back to its suction.

This is why we have balancing valves. They act sort of like police officers directing traffic. Basically, you throttle the balance valve on the circuit that's closest to the circulator more than you throttle the other balance valves. The greater pressure drop (Delta P) the water will experience as it tries to make it through that partially closed balance valve will encourage some of that water to take a different route, and that's just what you want.

Now there are a couple of ways you can set up your balance valves. You can try doing it by hand, which means you fiddle with this one over here and then that one over there until you

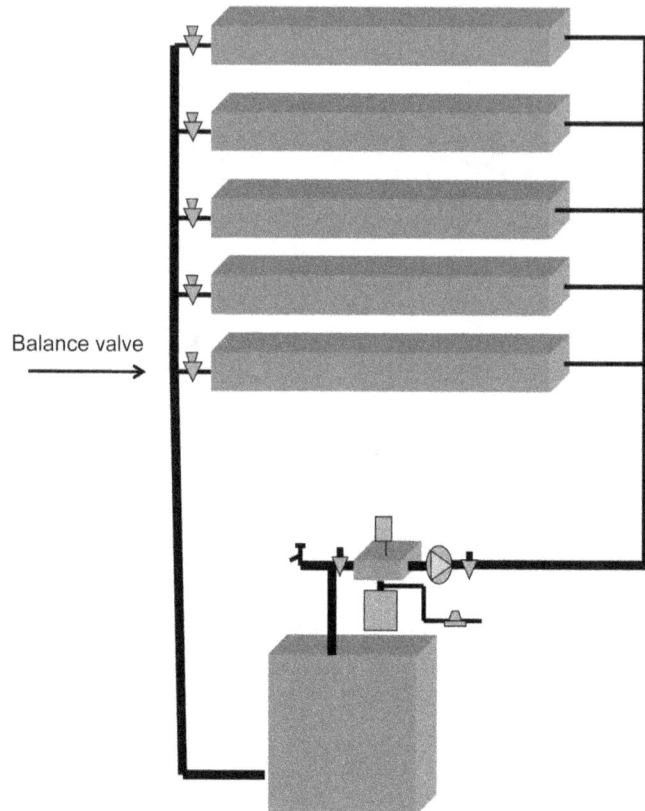

Balance valve

(hopefully) have everything just right. Or you can preplan the job on paper. The problem with the first method is that when you adjust one valve, you mess up what's going on with the other valves. It's like touching a mobile. The slightest touch over here makes the whole thing go out of balance.

True balance valves give you a way to preset each one to the flow rate that you need through each circuit (which assumes you actually know what that's supposed to be). You then fine-tune each of the valves to that setting in the field.

Here's a photo of this type of valve.

In classic hydronics, a lot of installers try to "balance" the system by oversizing the circulator. They figure that if they have enough pump power, the water is bound to flow everywhere, and given enough pump, the water probably will flow everywhere, but along with over-pumping comes velocity noise, and a big waste of electricity.

System balance is more of a ballet than a rugby match. Approach it with finesse.

Since they work by creating a pressure drop, it's best to install balance valves on the return side of whatever it is you're balancing. This is because when you drop the pressure on a fluid, gases will come out of solution. With classic hydronics, that fluid is water and the gas is air. Having the balance valve at the outlet side of the heater is especially important if you're working with coils. A balance valve on the supply side of the coil is liable to cause air to get trapped inside the coil, and that's never a good thing. Pipe it at the outlet. It's much better there.

Reverse-return

On smaller systems, where the main is no larger than two-inch, it often pays to use reverse-return piping rather than direct-return piping with balance valves. Here's a drawing of a reverse-return system.

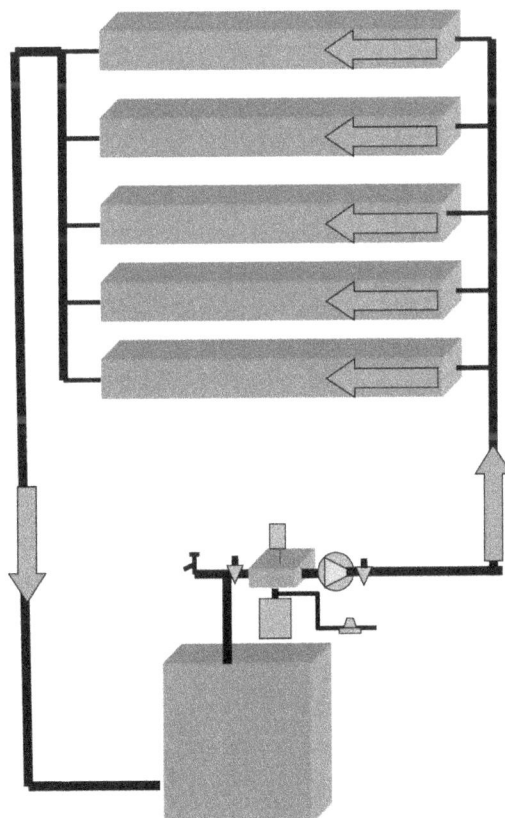

When we pipe this way, the distance (or Delta P) through each circuit is about the same because we've added that third pipe. Trace it through with the tip of your pen and you'll see what I mean. The path to, through, and back from every circuit is the same distance. There's no path of least resistance here, so the water flows evenly to all the circuits.

One thing to keep in mind, though, is reverse-return piping assumes that all your circuits have the same sort of radiation, and offer, more or less, the same pressure drop. You wouldn't mix, say, fan-coil units with baseboard in a reverse-return system because the pressure drop across the fan-coil unit will probably be greater than it is through the baseboard, and that would defeat the purpose of the reverse-return piping.

Just think like water and imagine your journey. You'll see it in your mind's eye.

CHAPTER EIGHT

Classic Circulators

In 1970, I went to work for a manufacturers' rep on Long Island and I fell in love with hydronics. We sold Bell & Gossett, and one of the older guys took me aside one day and taught me about the marble.

"When you're sizing circulators, or even when you're just looking at a hot-water job," he said, "you should imagine that you're inside the pipe, rolling from one end to the other like a marble."

I nodded and smiled. This was something that I could see in my mind's eye. "And whenever you get to a tee," he continued, "just ask yourself which way would you go if you were a marble."

"How will I know which way is right?" I asked.

"You'll *feel* it," he said.

He was teaching me the essence of Delta P that day. I liked his way because I have a good imagination and I really *could* feel it. Can you?

He also showed me how to use the Bell & Gossett System Syzer to size circulators, and that tool eventually became my American Express Card of Heating. I wouldn't leave home without it.

They had me working inside in those days. I took about a hundred contractor- and wholesaler calls each day, and many of those calls went like this:

"Hello, this is Dan."

"Yeah, whatever. Listen. I got this pump and it ain't working. I got to get a new one and I need it in a hurry."

"What can you tell me about the pump?"

"It's red."

"Uh huh, can you tell me anything else about it?"

"It's friggin' *BROKE!*"

And so we'd take it from there. I'd try to coax more information out of the guy. What was on the nameplate? Oh, there was no nameplate? How about on the motor? Could he get the horsepower rating? Can't read that either, eh? OK, how about the size of the flange? Or the length of the circulator from one end to the other? And the centerline between the flanges?

Often, I'd be able to look in the old B&G catalogs and find what I thought might be the proper circulator for the job, but this was making the big assumption that the circulator that was there now was the right one, and that was a big assumption to make. After all, the circulator was broken, and maybe the reason why it was broken was because it was the wrong size.

So I'd think about that marble. Which way would I go if I were rolling through the pipe? I liked to think of my marble as being large enough to touch the inside walls of the pipe. Big enough to feel the friction as its sides rubbed against the insides of the pipes. Use your imagination; that's the thing.

Can you feel it?

And there was also flow rate to consider, of course. I knew by this time that heat traveled on flow like a passenger on a train, and that the more flow there is, the more heat it can carry. You could see all of this on that beautifully simple tool, the System Syzer. It showed the normally used maximum flow rates for copper and iron pipes of various sizes. You could design for flows beyond these, of course, but you'd probably wind up with excessive pressure drop and velocity noise in most cases. If you stayed within the limits, you'd be safe.

For instance, the System Syzer tells me not to try to move more than four GPM through a 3/4-inch pipe. The folks who make copper, fin-tube baseboard show that same limit. That's good enough for me. If I needed to flow more than four GPM, I'd move up to a one-inch pipe, which can handle up to eight GPM.

When a contractor would call and tell me about the broken red circulator, and he didn't have any other information, I'd ask him what size pipe served the circulator. I'd figure that the circulator shouldn't try to pump more than that pipe could handle. Here are the limits I'd use:

1/2-inch pipe is good for 1-1/2 GPM.

3/4-inch pipe is good for 4 GPM.

1-inch pipe is good for 8 GPM.

1 1/4-inch pipe is good for 14 GPM.

1-1/2-inch pipe is good for 22 GPM.

2-inch pipe is good for 45 GPM.

And so on. If he could tell me the pipe size, I could tell him the flow rate.

Was I oversizing? Nope, I was sizing up to the maximum flow you could move through that pipe without having a problem. And that pipe was on the job.

Did the job need that much flow?

Can't tell for sure, but I was thinking that if I were the original design engineer, I would size for the smallest possible pipe to get the maximum flow needed, along with the best possible economy for the building owner. If the job needed 45 GPM, why would I use a main larger than two inches? It would be a waste of my client's money.

But pump head is another matter. I had to figure my total flow rate on the needs of all the circuits that the circulator would serve, but I would only have to base the pump head on what the flow would see while traveling through the highest-pressure-drop circuit. If my circulator was serving a two-pipe, direct-return system with five circuits, for instance, I would only have to worry about the worst-case scenario, which was usually the longest of the five circuits.

Direct-return, as we discussed, is sort of like a ladder. Figure one side of the ladder is the supply, and the other side is the return. The rungs are the radiator circuits. If I put in a circulator with enough head to make it up the supply side of the ladder, through the furthest rung on the ladder, and back down the return side of the ladder to the circulator's suction, then the circulator would have no trouble at all flowing through the lower rungs while it was at it.

It's sort of like saying that if you're able to drink six beers without falling down, then you'll also be able to drink fewer than six beers without falling down.

That analogy work for you?

Good.

So once we figured out the circulator's flow needs (based on the size of the pipe it served), I'd ask the contractor what the longest circuit was. Then I'd allow six feet of pump head for every 100 feet of length in that circuit. Again, this is from the System Syzer. When you're at the maximum, normally used hydronic flow through a pipe of any given size, the head loss is going to be four feet per 100 feet traveled. To this, we usually add 50 percent to allow for the additional pressure drop the water will see as it makes turns through fittings, valves and the other system components that create more friction loss than straight pipe does. If I add 50 percent to four feet of loss per 100 feet of travel, I wind up with six feet of head per 100 feet in that longest circuit.

Easy.

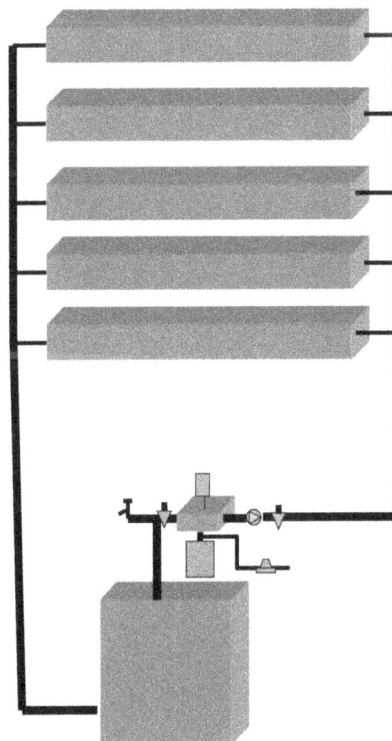

The contractor would tell me the length of the longest (highest-pressure-drop) circuit, and that would give me the two things I'd need to size the circulator over the phone - the flow and the head. If he didn't know the length of the longest circuit because all the pipes were hidden behind the walls and ceilings, I'd have him go outside and measure the length, width and height of the building. Then I'd imagine the longest run of pipe that I could put into that building.

I'd go from the left-hand corner of the building to the far end, then up to the top, across the top to the other side, down that side, back to the right-hand corner of the building, and finally, back to where I was standing. I'd base my pump head on that worst-case run, and then I'd select it from the pump curves in the manufacturer's catalog. And if my point of operation fell between a pump that was larger than what I had come up with, and another that was smaller than what I had come up with, I would always pick the smaller of the two.

I did this for the 19 years that I worked for that rep and I never once got a pump back.

Was I oversizing?

Nope. I was sizing based on the existing piping that was on the job. Someone once sized that pipe, based on the heat-loss of the building. I figured that if the pipes were the wrong size, someone probably would have noticed that long before the pump we were replacing had failed.

Today's smart circulators, the sort that run on ECM motors and magic, take all the guess-work out of circulator sizing. There's no longer a need to stand outside and measure the building. You don't even have to think like a marble. You just install that smart circulator and get out of its way. It will learn the system within no time at all.

I think these modern circulators are the berries, but they cost more than single-speed circulators, and you won't find them on every job, and certainly not on many of those classic systems. The old-school, pump-sizing method is simple, based on common sense, and it works. But it will cost more money to operate that classic circulator than it will cost to run a smart circulator.

As I said before, life is filled with choices. What's yours?

CHAPTER NINE

Classic Temperatures

Most classic-hydronic systems have the boiler running up to 180 degrees, with water returning from the system at 160 degrees. This rarely happens in real life, but it's the traditional way to design a classic hydronic system.

I wondered about this, and, having lots of time on my hands, I set out to find out how it all started. Why 180 degrees?

But before I tell you, I have to explain that I am the current custodian of a large collection of very old, heating-related books and magazines. Some of this stuff goes back to the 1700s. I often spend days in these old books. They smell like Mr. Hicks.

There are a lot of copies of *The Metal Worker*, a weekly trade journal that carried all the news of the day, both good and bad. I have every issue of that magazine for 1899, which was an important year in heating history, and a very wacky one as well. Many of the oddball devices that I've seen in basements over the years – the things that made me scratch my head in wonder – appeared as new products in that wonderful magazine during that fin de siècle year. Reading through those 52 issues is like getting into a time machine. And there's a certain peace to that. There is no uncertainty in spending time in the past. We know exactly what's coming next.

1899 was the year of The Carbon Club, an association of boiler manufacturers who got together in the spirit of what they called "cooperative competition," a concept that would send you to the slammer if you tried it with your competitors nowadays.

The Carbon Club guys were brazen in their very successful effort to control not only the price, but also the supply, of boilers. They did this for what they thought of as the good of the industry. They also did it because cast iron was scarce in 1899.

Oh, and they wanted to make huge profits.

What they did was in total violation of the Sherman Antitrust Act, which had been the law of the land for a decade. They didn't seem to care. They did what they wanted to do. It was a crazy time in America.

Here's a brief article from *The Metal Worker*'s July 15, 1899 issue. This is just nuts.

"A meeting of the Carbon Club will be held in New York on next Monday, when the reports of the Membership Committees will be made. While the desirability of securing all manufacturers as members is apparent, the condition of the iron market and the outlook is more important. The minimum price schedule adopted on June 20 has been found to contain some defects, and while these will be open for correction, it is probable also that a further advance will be made in the list. This, if done, will be due to the price and scarcity of iron, with the strong probability that the price of iron has not yet reached the top. Some manufacturers, both outside of the club and members, have arranged for their iron supply for the year, but as it could not be replaced except at the market prices, the price of boilers should be arranged in accordance or a radical advance would be necessary in boiler prices made by any manufacturer when his iron supply was exhausted and must be replaced at the much higher cost. This has been clearly pointed out, yet there are manufacturers who are holding to the former low prices, preferring to selfishly reap whatever benefit can be derived from being on the outside rather than to do their share to build up the market on a sound business basis. Should their example be generally followed, demoralization, and a year without a profit would result. Sometimes severe measures are necessary to open the eyes of the selfish. The Carbon Club is now strong enough to seek out the customers of such manufacturers and apportion them to the members with the instruction that prices must be quoted to them low enough to secure their trade. This would be drastic and not without its drawbacks, and it is to be hoped that cooperation can be secured by a more commendable method. Some members of the Carbon Club who were formally regarded as price cutters frankly state that though they suffered at the first advances they have now benefited by adhering to the course pursued and feel sure that others can be equally benefited by adopting the same course, whether members of the club or on the outside. The club, as far as can be learned, has been perfectly reasonable in all of its actions, and no considerable objection has been offered by the contracting trade."

Isn't that delicious? You're a selfish manufacturer if you lower your price to get business, and if you persist in that wackiness of trying to get more customers by dropping your price, the entire membership of The Carbon Club will come down on your head. They will seek out your customers and steal them from you, basically by giving away the boilers and driving you out of business.

How do you like *them* apples?

The "selfish" quickly got into line, and as I think about it, I realize that this was the boost the fledgling heating industry probably needed at the time. A very nice profit was guaranteed to all manufacturers, and the contractors went along with it. The building owners paid the quoted prices and central heating grew. All the boats rose with the tide, which The Carbon Club forced, but who knows how it would have gone, had they not done what they did? I don't approve, but I do understand. It was what it was.

This is the other thing that The Carbon Club did that year, and this forever changed the way we size hydronic heating systems. They waited until the very end of the century for this, and it makes me smile whenever I read it. Here you go. This was in the December 23, 1899 issue of *The Metal Worker*:

"A meeting was held of the Carbon Club at the Murray Hill Hotel, New York, December 18 and 19, with a large attendance of the members. Several applications were received from manufacturers and some new members were elected. The recommendations of the Committee of Boiler Ratings, which were discussed at the November meeting, were taken up, and after some minor changes, were adopted. This is virtually a standardization of the home heating boilers made by the members of the club, and with the uniform rating and uniform prices many of the perplexities of the trade are removed. All boilers are now rated on a proportion of 100 for steam and 165 for water, with steam at 2 pounds pressure or water at 180 degrees at the boiler. The rating now includes all mains, returns and risers as heating surface, and the surface exposed in them must be added to the surface required in the radiators to determine the boiler power needed. It is only necessary for the trade to understand that the mains must be considered to avoid purchasing a boiler that is too small. If a boiler shows the 2 pounds steam pressure or 180 degrees temperature in the main when at work, the rating will be considered verified by the manufacturers. The new list also divides boilers into two classes. A uniform rating has been agreed upon for tank heaters on a basis that they will heat 130 gallons of water for every 100 feet of surface that they are rated to carry, and their prices have been rearranged so that concessions are made to the buyer on some sizes."

What we have here is the agreement between all the boiler manufacturers of the time that a hot-water boiler be proportionally larger by 65% than a steam boiler serving the same building. You see that today when you look at the difference in the value of Equivalent Direct Radiation for steam and hot water. With steam, a radiator will put out 240 Btuh per Sq. Ft. EDR, but that same radiator, filled with hot water, will put out only 150 Btuh.

They also agreed that no steam heating system from that day forward should need more than 2-psi pressure at the boiler to heat the building. This was a very significant decision because it put a stop to what was becoming a very dangerous situation. Contractors had been using boiler pressure as a competitive edge. They were sizing systems with as much as 60-psi pressure at the radiators. Higher pressure means smaller radiators and pipes, but the problem was (and remains) that all steam-heating systems have to start at 0-psi pressure, and at the lower pressure, the steam can (and did) suck the water out of the boiler (low-pressure steam moves faster than high-pressure steam). This caused boilers to either dry-fire or explode.

The Carbon Club put a stop to that wackiness. They did it by establishing standards for sizing steam pipes. These standards allowed for one-ounce of pressure drop over 100 feet of travel. They leveled the playing field for the contractors, and it's the reason why they can heat the Empire State Building on most days with just 1-1/2-psi steam pressure.

At that meeting, the members of the club also recognized that there is a piping pick-up factor, which must be recognized by contractors when they size a boiler, lest they undersize a boiler, which would be very bad for the manufacturers.

They established a standard for heating domestic hot water, which stood for many years, and finally, they let the proof be in the pudding. If a contractor could heat the entire building with a boiler that contains no more than 2-psi steam pressure, or 180° hot water, then he had picked the right boiler for the job. If it couldn't do it at that pressure or temperature, then he had screwed up somewhere and the problem was on him.

And that's why we use 180 degrees as a benchmark high-limit setting.

And here's why we use 160 degrees as the return temperature

This goes back to the days before we had electronic calculators, and I suspect that's the reason why we inherited the 20-degree temperature drop in hydronic systems. We start with this formula for figuring how many gallons we need to flow through the system each minute. Here 'tis:

$$GPM = \frac{Btuh}{\Delta T \times 500}$$

Let me explain what the parts of the equation mean. Let's get the simple one out of the way first. GPM stands for Gallons Per Minute. It's the standard term that we use in North America when we're talking about hydronic heating systems. We move so many gallons of water per minute, and the heat travels on that flow like a passenger on a train. GPM is an expression of volume over time. It has nothing to do with the speed of the water (that's velocity). GPM is all about the *amount* of water that can move from here to there during one minute. How fast that water moves depends on the size of the pipe through which it's moving, and the circulator that's providing the force. In other words, 10 GPM moves faster in a one-inch pipe than it does in a three-inch pipe, but it's still 10 GPM, regardless of the size of the pipe.

Okay, the next component is Btuh. That stands for British thermal units per hour. And in case you don't know, a British Thermal Unit is the amount of heat it takes to raise one pound of water (that's about one pint) one degree Fahrenheit.

And by the way, this wasn't always the case. Thomas Tredgold, the Englishman who came up with the British thermal unit in the early 1800s, first decided to make it the amount of heat required to raise the temperature of **one cubic foot** of water one degree Fahrenheit.

He just made that up.

They changed it after he died. He wasn't able to argue.

So there.

The triangle symbol in the equation is the Greek letter Delta. We've met that before. Delta (or Δ) means "Difference in." So Delta T (or ΔT) means "Difference in Temperature. And as you know, Delta P (or ΔP) means difference in pressure.

When we're talking ΔT we're referring to the difference in temperature between the water that leaves a boiler and the water that returns to the boiler after it's been through radiators, convectors, radiant panels, or whatever.

Okay, last item in the equation - the number 500. This is actually a shortcut. Since we're using this equation to figure out how much water we have to move in one minute (GPM), we have to know something about the water itself. In this case, we need to know that a gallon (as in GALLONS per minute) weighs 8.33 pounds. The other part of the 500 has to do with time (as in gallons per MINUTE). There are, as you know, 60 minutes in an hour. So if you multiply the weight of a gallon of water (8.33) by 60 minutes, you'll come up with 499.8, which we'll round up to 500 to make life easier, and that's where we get that part of the equation.

Pulling it all together, we have a certain amount of heat leaving the boiler, which is the top part of the equation, and we're using that heat to raise a quantity of water (measured in pounds) a certain amount of degrees Fahrenheit over the course of one minute. Knowing all that, we can figure the flow rate. Make sense?

Okay, here's why we work with that 20-degree temperature difference across the typical hydronic system (even though it rarely occurs in real life). If you replace the ΔT symbol in the formula with the number 20 (as in a 20-degree temperature difference), the bottom part of the equation will now look like this:

20 X 500

Remember, they came up with this in the days before calculators. Multiplication such as this you can do in your head. Multiply 20 by 500 and you get 10,000, right? Now all you have to do is divide the Btuh load by 10,000 to get the required GPM for the system. For instance, if you need to move 100,000 Btuh from a boiler to a bunch of radiators, all you have to do is size a circulator for 10 GPM (100,000 ÷ 10,000 = 10). See how simple life can be? The water goes out into the world and is supposed to return 20 degrees cooler.

But that doesn't mean you have to stick with a 20-degree temperature difference. There are times when a wider temperature difference works to your advantage, particularly if you're using that pumping technique called primary-secondary. With a system piped that way, you might want to work with a 40-degree temperature drop instead of a 20-degree temperature drop.

Take a look at what that does for this equation:

$$\frac{100,000 \text{ BTUH}}{\Delta T \text{ X } 500} = GPM$$

Substitute 40 for ΔT and look at what happens:

$$\frac{100,000 \text{ BTUH}}{40 \text{ X } 500} = GPM$$

Solved, it looks like this:

$$\frac{100,000 \text{ BTUH}}{20,000} = 5 \text{ GPM}$$

See? You just cut the required flow rate in half. You also reduced the size of the pipe, and perhaps the size of the circulator. That saves money.

If you want to see how this works on a bigger job, just add zeros.

Engineers will play with Delta-T on systems that have primary-secondary pumping, and they'll use even wider temperature differences for the secondary circuits that feed radiant systems. I learned this the hard way.

A bunch of years ago, when Al Levi was still in his family's heating business, they decided to move to a new building here on Long Island. It was one of those rare, down-to-the-bare-walls opportunities, where the owners of a very good heating company get to decide how to heat their own building. From what I've seen, most owners take this opportunity to be frugal by selecting the cheapest system they can conjure. Then they set out to sell their customers golden boilers and platinum pipes. Most contractors don't walk it like they talk it, but the Levis aren't most contractors.

Al and his brothers came to me and asked if I would think with them about how to best heat their new building so that they could also use it as a showroom for heating possibilities. I came up with this primary-secondary pumping system that incorporated two, staged boilers, and more than a dozen zones – some radiant, some high-temp, and one indirect domestic hot-water tank.

We started with a big rectangle of rather large pipe that hung from the ceiling and hugged the perimeter of the building. From this, we hung the zones, each with a small water-lubricated circulator. We piped the boilers the same way. The large perimeter loop had its own circulator and it was the primary loop. Everything else (the radiators, the radiant, the indirect tank, and the boilers) were secondary circuits. The radiant used a three-way valve to lower the supply temperature to what it needed.

We started the system and it ran pretty well, although there was some creeping gravity flow down into some of the secondary circuits when they were off. The guys added flow-control valves to the supply and return of each secondary circuit and that solved that problem.

Years later, the system is still running well, but whenever I stop by their place to visit, I look things over and wish I had a second chance at this one. I think I could do it better now.

What we built back then was what many others built, and continue to build. It was a one-pipe, primary-secondary system, which seems pretty easy when you first look at it.

We based the whole thing on a 20-degree temperature drop and that gave us our fairly large, primary-circuit pipe. It wasn't until years later that I realized that a 40-degree temperature drop across the primary should be the norm, not a 20-degree drop. Most of these boilers fire up to 180 degrees because of the needs of the indirect, domestic-hot-water tank. Any boiler

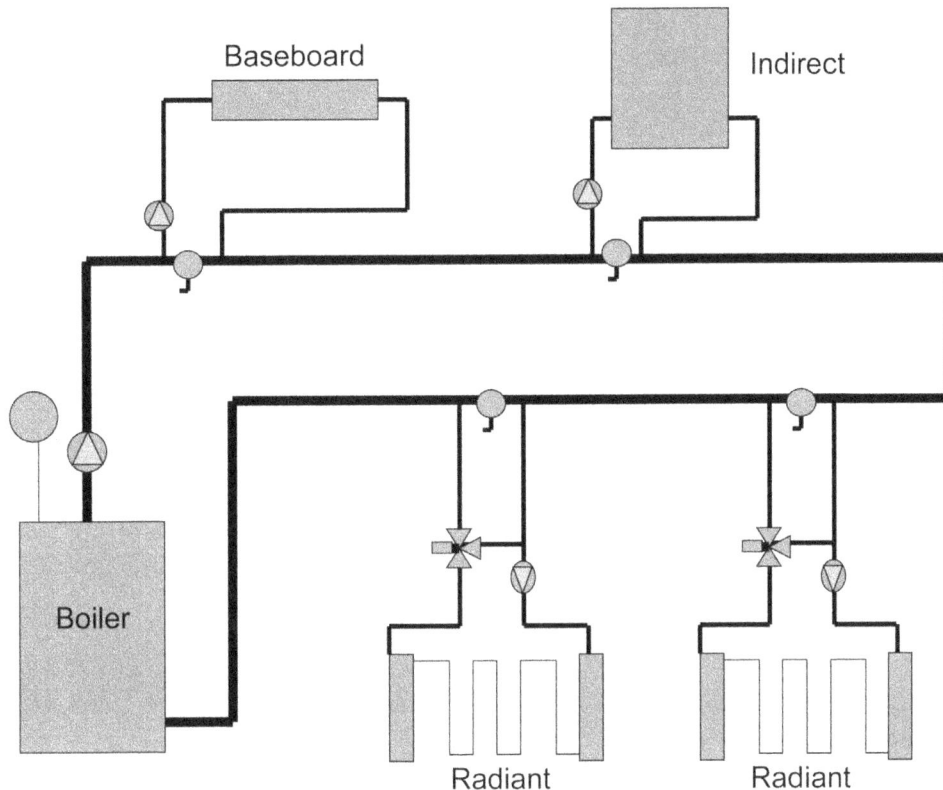

can tolerate return temperatures down to 140 degrees F without condensing flue gases, so why the heck was I designing my primary loop for a 20-degree temperature drop back then?

It was a habit. I did what I was raised to do. With a 20-degree drop, the math is easy. It's a safe and traditional way to get the job done. If something doesn't work out, you can always bump up the aquastat. Hey, that's what Grandpa did.

The Levi's system worked okay with that 20-degree drop. What nags me is that it could have worked just as well, and at a lower initial cost, if I had gone with the wider temperature drop, the smaller primary loop, and the smaller primary pump. And even though it wasn't my money, it bothers me. My old habits cost my friends money.

And then there's this other thing: We had more than a dozen zones hanging from that primary main, and many of them needed 180-degree water on the design day. There was the indirect, domestic-hot-water tank, along with some fan coils, cast-iron radiation, panel radiators, baseboard convectors, and a few other high-temp units. But since this was a one-pipe, primary-secondary system, there was no way that all of those units could possibly get 180-degree water on the design day. Return water from each circuit was blending back into the primary circuit, lowering the temperature of the supply water to the following circuits. It follows this formula:

(Supply GPM X Supply Temp.) + (Return GPM X Return Temp.) = (Mixed GPM X Mixed Temp.)

The result was that, on the design day, they had to raise the high-limit temperature above what it might have been, had I used a better design. My piping choice was raising their fuel usage.

A few years ago, I went back to the manuals that Gil Carlson had left us, and I spent a summer studying them once again. This time, I tried to read them through the perspective of how designers and installers did primary-secondary pumping in the '60s, and how they do them today. What came out of that summer was a book I called, **Primary-Secondary Pumping Made Easy!** You can find it in the store at HeatingHelp.com. That book was my attempt to bring Gil back for one more lesson. It's my humble words and observations, but the main thoughts and calculations are pure Gil. I just took another look at the topic through the prism of nearly 60 years. I didn't expect that many readers would sit and do the math on every job, but what I hoped with that book was that my reader would come away convinced of at least two things:

First, that it makes no sense to design primary-secondary piping systems around a 20-degree drop (if there's radiant involved, that's probably going to be seeing an 80-degree temperature drop).

Second, that when it comes to one-pipe, primary-secondary systems it really pays to use supply-and-return manifolds, grouped to the secondary circuits, and based on required supply temperature. Like this:

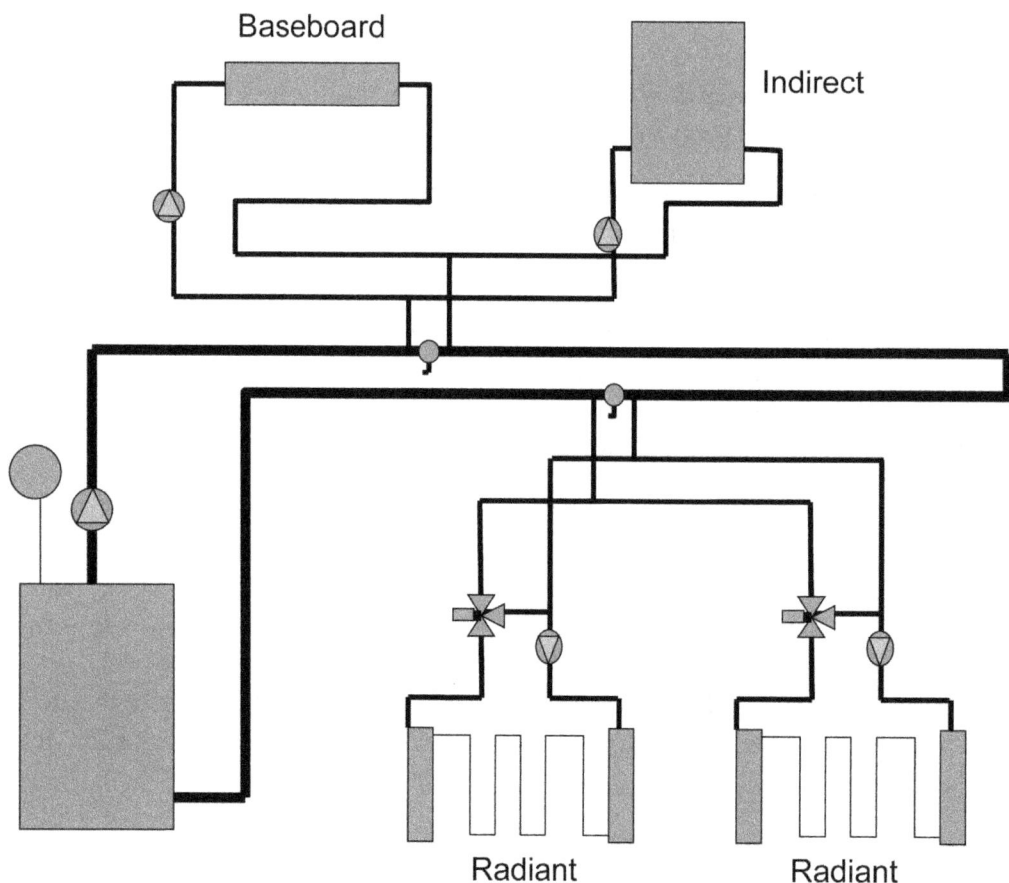

The manifolds give you the same advantage that you have with a two-pipe primary-secondary system (or any two-pipe system) in that you can be certain that if two or more secondary circuits need, say, 180-degree supply water, they will get that temperature. In the book, I ran an example of two, nearly identical, high- and low-temperature systems. The only difference was that one system had manifolds and the other system was similar to the one I had designed for the Levis, where each secondary circuit drops independently off the primary. The system with the manifolds ran 10 degrees cooler on the supply temperature. Same level of comfort, but at a lower temperature, which means less fuel burned. Nice.

You set up the supply and return manifolds off the primary loop by spacing the tees to the common piping as closely as possible. Group your high-temperature secondary circuits on one manifold and your low-temperature circuits on another. The system will be much more predictable, and it will burn less fuel.

So use manifolds and wider temperature drops on those primary-secondary systems. You won't get into trouble, and your clients will save on both the initial costs, and on fuel as the years go by.

It's a classic.

CHAPTER TEN

Classic Air Problems

The first time I saw a Spirovent air separator in action was at a plumbing supply house in Brooklyn, New York. It was set up in this rig that had see-through piping and it was drawing lots of attention. I got to use a bicycle pump to inject air into the moving water and that was fun. I gave it a good shot of air and watched as this new type of air separator caught the bubbles and spit them out.

I'm easily amused.

The salesman smiled at me and told me about microbubbles (a new word for me at the time). He explained how the microbubbles would collide with, and adhere to, the inner workings of the Spirovent, and then leave the system by way of the built-in air vent. He also mentioned Henry's Law, which was a law I had never heard of, but I nodded with great solemnity. I didn't want him to think I was dumb.

When I got home, I looked up Henry's Law in the dictionary and learned that it had to do with the way a gas will dissolve in a liquid, depending on the pressure applied to that liquid. It turns out the more pressure you put on a liquid, the more gas it will hold in solution. Vice versa, of course.

Oh, and the hotter the water gets, the less air it will hold in solution (and vice versa). I sort of knew all of that by watching water boil, ice freeze, and by shaking bottles of soda pop, but I had never put a name to it. Now I knew. It's all about William Henry.

So here's his story, and it's a grand one.

William Henry's daddy, Tom, was a rich doctor, who also owned an industrial chemical business in Manchester, England during the late 1700s. Try putting those two professions together nowadays. You could probably make your own patients.

Tom Henry was the first guy to suggest that you could bleach clothing with chlorine. How about that? Isn't it good to know that Clorox and the proper location for a hydronic air separator have their roots within the same family?

Anyway, Little Billy Henry showed up just before the Christmas of 1774 and all went well until he reached the age of 10. That's when a beam fell from the ceiling, landed right on him, and left him with chronic pain for the rest of his life. Because of this, he wasn't able to play with the other kids, so he stayed inside and hit the books.

Oh, and he hit them *hard*.

At age 16, he began studying medicine, and at 21 he entered the University of Edinburgh in Scotland, but only stayed a year. He left the university to help his father with his medical practice and to work in the family business. He spent the next 10 years doing original research in chemistry (which plays a big part in this story), and then, because he didn't like to leave things undone, he returned to medical school at age 31 and got his M.D. two years later.

Oh, and he did his dissertation on, of all things, uric acid (a.k.a. pee), which I think is splendid because it ties Bill Henry even closer to the business of plumbing and heating.

Now we are all influenced by those who came before, and Bill Henry was no different. A generation earlier there had been a fellow in France by the name of Antoine Lavoisier. Ever hear of him? He was the guy who first said that matter can neither be created nor destroyed, and he also gave names to two things that are pretty important – those being oxygen and hydrogen. Lavoisier also came up with the first extensive list of elements, and helped create the metric system.

So there. What have *you* done today?

And you would think that the French would have appreciated all of this, but at the height of the French Revolution, someone accused Antoine Lavoisier of selling watered-down tobacco, so they gave him a very quick trial, and then cut off his head.

Smoke 'em if you got 'em.

Bill Henry was fascinated by Lavoisier's work, and in 1801, while still working with his dad, he put together a book about it, and he did a fine job of explaining it all. He called the book, *Elements of Experimental Chemistry*, and this went through 11 editions over the following 30 years. He kept adding to it, and it was this work that introduced generations of chemists to the Frenchman Lavoisier's careful use of experimental measurement. Bill Henry was just 27 years old when he wrote that book, and two years later he published the paper that established what we now call Henry's Law. Years and years after that, I'm in some plumbing supply house in Brooklyn, NY, and an air-separator salesman is bringing it to my attention.

Ain't life grand?

And how about this? Henry's Law came about because Bill Henry was sitting around wondering why our atmosphere, which is composed of all these different gases, each with its own density, doesn't separate into layers, as oil and water will separate when put into the same container.

I've never wondered about that.

Have you?

I mean I've lived more than six decades on this planet without *once* considering that.

Gosh.

Anyway, it was Bill Henry's initial thinking about our mixed-up atmosphere that lead to the theory of mixed gases, which we today credit mostly to Bill's pal, John Dalton.

I have to tell you about him too. You've probably never heard of him, right? I hadn't. He's the guy who figured out that atoms make up everything.

And consider this for a moment: These two guys were hanging out together. One's figuring out atoms and the other's wondering why all the atmosphere doesn't look like a seven-layer cake. They're buddies.

Isn't that something?

Dalton was brilliant, but he was clumsy and careless around the lab. He also had very little money for experimenting (he worked as a teacher). His buddy, Bill Henry, however, had lots of money, and even more patience, so the two men worked together like oxygen and hydrogen.

They did most of their experiments with gases because gases are chemically simpler than other forms of matter, and when you're looking for something as small as an atom, keeping it simple really helps. Out of all this experimenting came Henry's Law.

Think about that the next time you're installing an air separator, or wondering why the air keeps coming out of that compression tank. Somebody before us had to figure out all of this stuff.

John Dalton goes on to become famous for the atomic theory, which is very cool because suddenly, the world could identify and order elements. From this comes the Periodic Table of the Elements, something you probably had to study in school at some point. Ugh.

John Dalton and Bill Henry are also the guys who gave us the term, H_2O. There's something to consider the next time you're purging those pipes.

And how's this for being wonderfully human? Although Bill Henry's experimenting helped John Dalton come up with atomic theory, Bill didn't want to back it. As he got older, he became more reluctant to accept change. He didn't like it when his experiments pointed to something other than what he expected. He held onto his old beliefs, such as insisting that heat has mass. (It doesn't.)

Human nature.

In 1824, a series of unsuccessful surgeries on his hands took away his ability to manipulate instruments. He quit chemistry and turned his full attention to medicine, specifically to the spread of contagious diseases.

In 1831, a cholera epidemic hit the United Kingdom and it was horrible. Nowadays, we know that the way to prevent the spread of cholera is to wash our dirty clothing in real hot water and chlorine bleach, which Bill's father, Tom, had promoted years earlier, but they didn't try chlorine then. Instead, Bill came up with an inexpensive and simple device that used heat to disinfect clothing. It worked, and it probably would have saved countless lives, but for some reason, Bill Henry decided that he didn't like the idea of the device and so he abandoned it.

Thirty years later, Louis Pasteur came up with the germ theory of disease and we all began to pasteurize things with heat. How many people must have died needlessly during those 30 years because Bill Henry changed his mind?

In 1836, chronically depressed and filled with the pain that had been with him since that long-ago childhood accident, Bill Henry took his own life.

And I hope you'll never look at those air separators, air vents, and compression tanks in quite the same way ever again.

A bit of chemistry

Air is mainly nitrogen (80%) and oxygen (20%), which is a good thing for us. In a classic hydronic system, there's a lot of iron and steel, and the oxygen in the water will react with that metal and form rust. That leaves us with lots of nitrogen, which will dissolve in the water and come out of solution when the water gets hot, or the pressure gets low (that's Henry's Law).

Here's a chart that shows you how much nitrogen will dissolve in water, depending on the pressure and temperature.

From this chart, you can see why most air separators wind up very near the outlet of the boiler. The hotter the water gets, the less nitrogen (and how about if we just call it air?) you'll have in the water. It's best to grab the bubbles right there at the hot spot.

An alternative place for the air separator would be at the top of the system because that's where the static pressure is the lowest (again, Henry's Law). The trouble with this, though, is that you may need more than one air separator because there may be more than one high point in the system. I also think it's best to grab the bubbles at the boiler before they have a chance to enter the system, but it's good to have options. If you want to put air separators

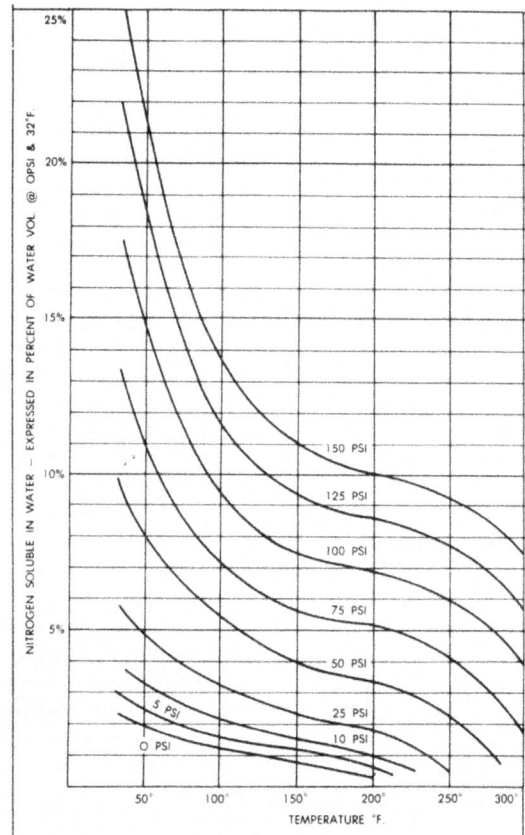

all over the system, that's fine with me, and I'm sure the air-separator manufacturers will appreciate it.

And while we're talking chemistry, let me put in a plug for cleaning the inside of those classic hydronic systems after you work on them. There are a number of companies that offer cleaning chemicals made specifically for hydronic systems, and it's a great idea to use these because of all the oil that goes into cutting and threading pipe, and all the solder and flux that's involved with connecting copper tubing to fittings and valves.

The oxygen in the fill water will react with the oil and any other foreign matter you leave inside the system and that will lower the pH of the water. Ideally, the pH should be slightly alkaline; somewhere between 7 and 9 on the pH scale is just right. If the water gets acidic, it will corrode the metal and that's never good. The air in a system that's acidic will often burn when you vent it, so that's an easy way to check to see what's going on. Hold a match near the air vent. You just may get a surprise.

Be the air

You're air and you're lighter than water so you're going to rise to the top. That's why the Dead Men put their expansion tanks at the top of those gravity systems. When they moved the tanks to the basement to keep them from freezing, those expansion tanks became compression tanks. They had to put the air in there, and more often than not, the air left the tank by gravity circulation. I'll tell you more about that in a minute, but first, let's take a look at the problems that too much air in a hydronic system can cause.

First, if you're the air, you have the ability to be noisy as you rattle through the pipes. And if the system isn't well balanced, you're liable to make waterfall noises as you tumble off a horizontal pipe and into one that heads downward. You're annoying people, and that's going to cause problems. Someone's going to have to show up and evict you.

When air gets trapped in a radiator, it can cut down on the radiator's efficiency by limiting the water's available contact area with the metal. In some cases, air can stop the flow through the radiator, and where there is no flow, there is no heat.

Troublemaker!

If you, as the air, get trapped inside a water-lubricated circulator, you can really stop the flow, especially if you get trapped between the impeller and one of those built-in, flow-check valves that the circulator manufacturers offer nowadays. When that happens, the circulator will essentially be running dry and it won't last long. You're a killer.

And the more there is of you in the system, the more your oxygen component is going to want to react with the cast iron and steel in the system, so you'll be making rust, and rust can clog up the works. Nasty!

And you don't have to be big to cause problems. You can be microscopic within the flow of the water. We'll call you microbubbles.

I was at the airport with The Lovely Marianne, getting ready to board a plane. I have a problem paying four bucks for a bottle of water, so when I travel, I always bring an empty bottle from home and fill it from a drinking fountain once I clear security. On this occasion, I filled the bottle and it looked like milk. The Lovely Marianne scrunched up her pretty face and said, "Eww, I'm not drinking that!"

"But it's just microbubbles, my sweet," I explained. "If we had a microbubble reabsorber, we could clear them right now, but since we don't, we just have to wait a bit for the microbubbles to settle out of solution, rise to the top and leave. I love you, and I would *never* let microbubbles touch your precious lips."

"You're ridiculous," she said.

The woman is crazy about me.

In a closed hydronic system, those microbubbles get together like a flash mob and create bigger bubbles, and those are the ones that will ruin your day. It's best to catch those tiny bubbles (think Don Ho) while you can and spit them out. More on this in a little while, but first, let's consider why air leaves a compression tank.

Remember this drawing from earlier when we talked about Gravity systems?

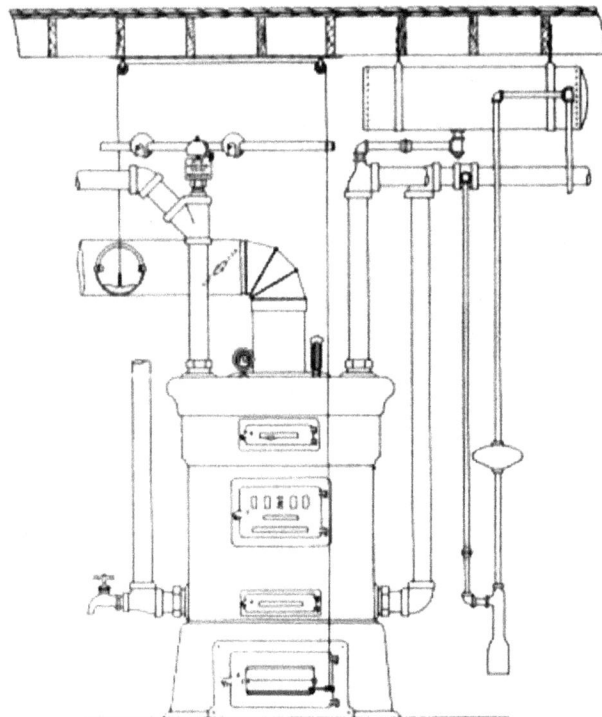

They moved the tank to the basement to keep it from freezing, but look at the way they connected the tank to the system. It took them a long time to figure this out, but when you connect a plain compression tank (that's one without a diaphragm), the water in the tank will absorb the air in the tank. Then, gravity circulation within the line that connects the tank to the system will move the cooler, air-laden water out of the tank. It does this because the water in the tank is cooler than the water in the pipes. Hot water rises; cold water sinks, and you don't need two pipes to make this happen. One will do. The rising and falling waters pass each other within that single pipe.

And why does the cooler tank water absorb the air sitting on top of it?

Henry's Law! (He just keeps popping up all over the place, doesn't he?)

Once that air-laden water leaves the tank and merges with the hot water in the pipe, the air comes out of solution and the bubbles start looking for other bubbles. They love company.

If you vent them from the system, or if an automatic air vent vents them from the system, the system pressure will drop. The feed valve will then open to add water (assuming you've left the feed valve in the open position). And since the pipes and radiators are already filled with water, the only place that new water can go will be into the compression tank. And that's why compression tanks that are hooked directly to the piping will waterlog.

It's a real pain.

To solve that problem, hydronic-equipment manufacturers came up with a device that stops the gravity circulation between the compression tank and the system piping. Here's an example of one of those devices.

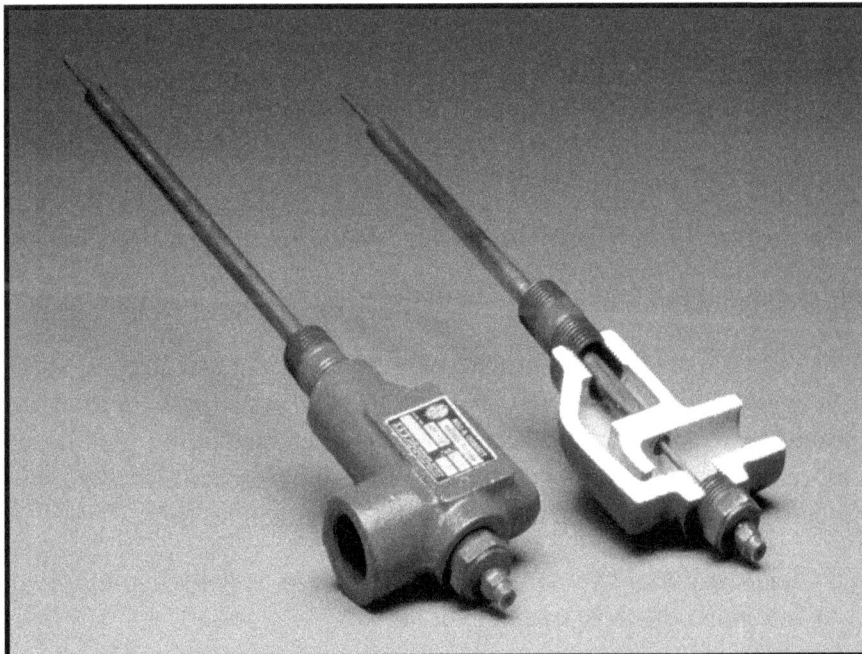

Bell & Gossett makes this one and they call it an Airtrol Tank Fitting. I like this device because it has no moving parts. It works by following the laws of physics. When you trap air bubbles with an air separator, they rise by their buoyancy and enter the Airtrol Tank Fitting. They go up that tube and enter the air cushion. Here's a drawing of that happening.

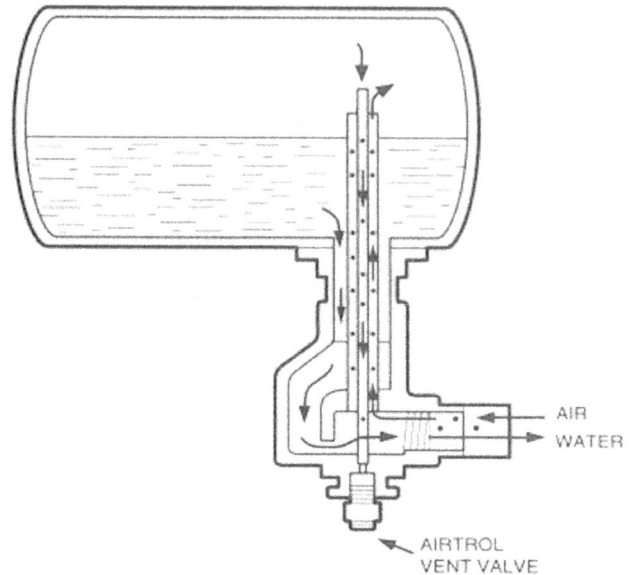

The tube is higher than the water level in the tank. It's sort of like a hydronic periscope. You set that level when you first fill the tank by loosening that fitting at the bottom of the tube. The tube should be two-thirds the diameter of the tank, and oh, always put the air at the *top* of the tank. It takes too long to try to do it any other way.

It does!

Anyway, when hot water from the heating pipe rises up the line to the tank, it can't sneak under that baffle that's near the bottom of the Airtrol Tank Fitting. Hot water rises; it doesn't sink. So it goes up the tube, but the tube is higher than the water level in the tank, so when the hot water gets to the top of the tube it stops rising. And if the hot water can't get into the tank, the cooler water that has absorbed some of the tank air won't fall out.

You just solved the gravity-circulation problem.

While we're looking at those old tanks, I want to mention what goes into sizing a compression tank. You have to consider these things:

1. The amount of water in the whole system

2. The difference in pressure between the feed-valve setting and the relief-valve setting

3. And the average water temperature of the system on the coldest day of the year.

When you have a system with a lot of water in it, such as an old gravity system, where both the pipes and the radiators are large, you're going to wind up with a compression tank that's larger than it would be on a modern system, which contains much less water. If you're using an old-school compression tank without a diaphragm on that old system, you can start by measuring the total system equivalent direct radiation (EDR) and then do this:

1. When there is less than 1,000 square feet of radiation on the job, multiply the total EDR by .03 to determine the tank size in gallons.

2. If the total EDR is between 1,000 and 2,000 square feet, use .025 as a multiplier.

3. If the total EDR load is greater than 2,000 square feet, use .02 as a multiplier.

More water means more expansion, and that's why the tank will be so big. If you decide to use a diaphragm-type compression tank on that job, it's also going to be bigger than it would be if it were on a modern system. Probably a lot bigger.

Here's a rule of thumb for the diaphragm tank:

Take the size of the standard steel compression tank in gallons and multiply by .55 if the building is two-stories tall or .44 if the building is three-stories tall. The answer will give you the *volume* of the diaphragm tank.

Here's an example. Let's say we have a two-story house with 1,000 square feet of radiation. We'll size a standard steel tank first: 1,000 X .03 = 30 gallons. Now, since it's a two-story house, we have to multiply that by .55 to get the volume of the diaphragm tank. (30 X .55 = 16.5 gallons of required volume in the diaphragm tank)

You can find the volume of the diaphragm tank in the tank manufacturer's specification sheet. Here, for instance, are the volume ratings of standard diaphragm-type tanks made by Amtrol, Inc.:

Model Number (Amtrol)	Volume (in gallons)
15	2.0
30	4.4
60	7.6
90	14.0
SX-30V	14.0
SX-40V	20.0
SX-60V	32.0
SX-90V	44.0
SX-110V	62.0
SX-160V	86.0

And here are the volumes of the tanks made by Vent-Rite (Flexcon Industries):

Model Number (Vent-Rite)	Volume (in gallons)
VR 15 F	2.1
VR 30 F	4.5
VR 60 F	6.1
VR 90 F	21.0
SX VR30 F	21.0
SX VR40 F	21.0
SX VR60 F	29.0
SX VR90 F	37.0
SX VR110 F	53.0
SX VR160 F	74.0

For the building in our example, you'd use an Amtrol SX-40-V, a Vent-Rite VR 90 F, or any combination of smaller tanks that will equal or exceed 16.5 gallons of volume. If you wanted, you could use four Amtrol 30s or four Vent-Rite VR 30 Fs, for instance.

Oh, and before you install the tank (or tanks), check the air pressure on the diaphragm side of the tank. As we discussed, it should equal the system fill pressure when you have the tank disconnected from the system. The fill pressure for a two-story building is typically 12 psig; for a three-story building, it's 18 psig. If the pressure is too low, use a bicycle pump or an air compressor to increase it. The tank's pressure (when disconnected from the system) should *always* equal the system fill pressure (the pressure reducing valve's setting).

I know I'm repeating myself there, but it's important.

Okay, with that out of the way, let's talk about getting rid of the free air when you're first setting up the system. Once you heat the water, more air will come out of solution, and getting rid of that is going to be the air separator's job. We'll get to those in just a bit.

Purging free air

The old-school way of getting rid of the free air that's everywhere inside the system when you first start it, or after you've drained it, is to use a hose at the end of the zone you're purging, and to open the fill valve, or better yet, to use a bypass around the fill valve (or the fast-fill feature on the fill valve). You need full volume to get a good purge, and the flow through a feed valve usually isn't enough, especially if the pipes are large. If all goes well, most of the air will scoot ahead of the water that's flowing into the system and leave through the hose.

A great way to pipe if you want to get rid of the free air in a hurry is to use the Pumping Away techniques that we looked at earlier. Remember?

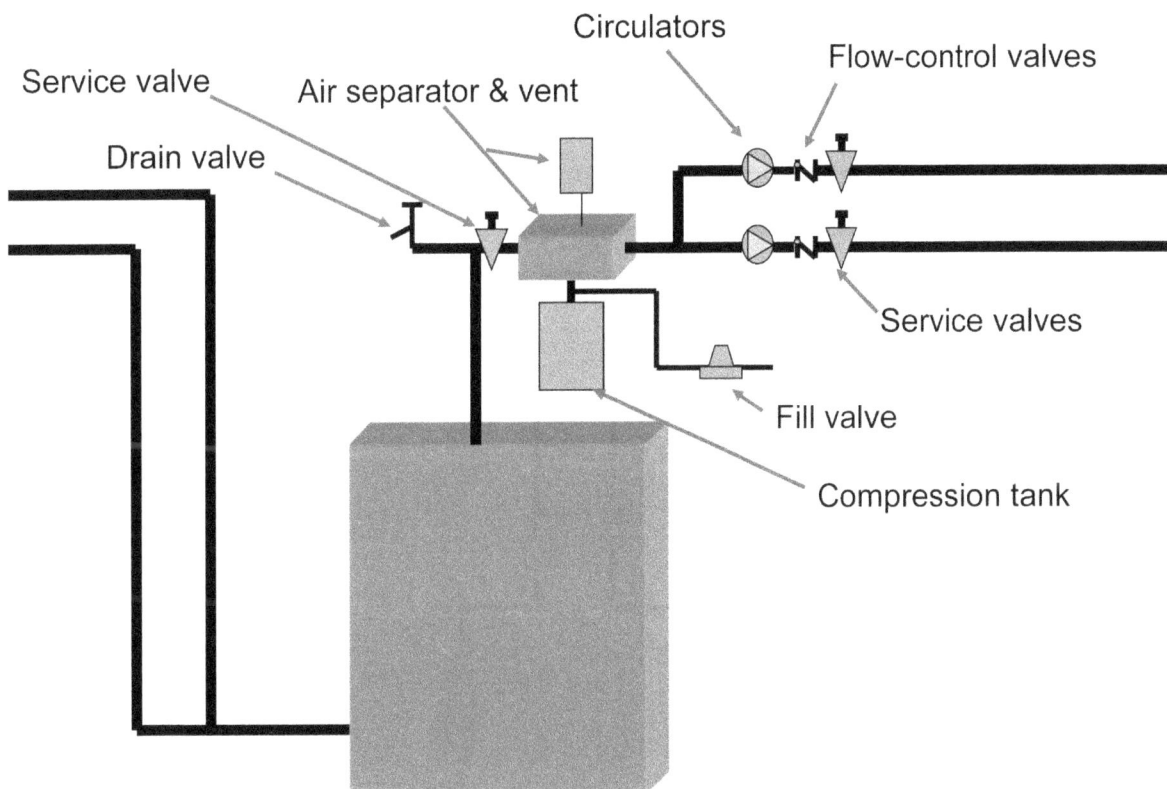

This is easier to start than a system with the purge valves on the return side. Just shut that main service valve, the one that's right at the outlet from the boiler. Close the valves to all the zones and open the feed valve's fast-fill, or a bypass line. Then open one zone at a time. The water will force the air through the system, into the bottom of the boiler and out the hose that's just before that main service valve. Then do the next zone, and so on. There are fewer places for the air to hide, and when you start the circulator, it will deliver its full differential pressure to the remaining air in the system. The bubbles will be smaller and easier to move.

Primary-Secondary pumping and piping is another technique you'll see on both one- and two-pipe systems. With these systems, the two, closely spaced tees separate the secondary circuits from the primary circuit. By having very little common piping between the two circuits, circulators of different sizes can work together and get along, without one circulator back-pressuring another.

But because of those closely spaced tees, purging can be a problem, so it helps to install a full-port ball valve on the primary pipe and between the two tees. Like this:

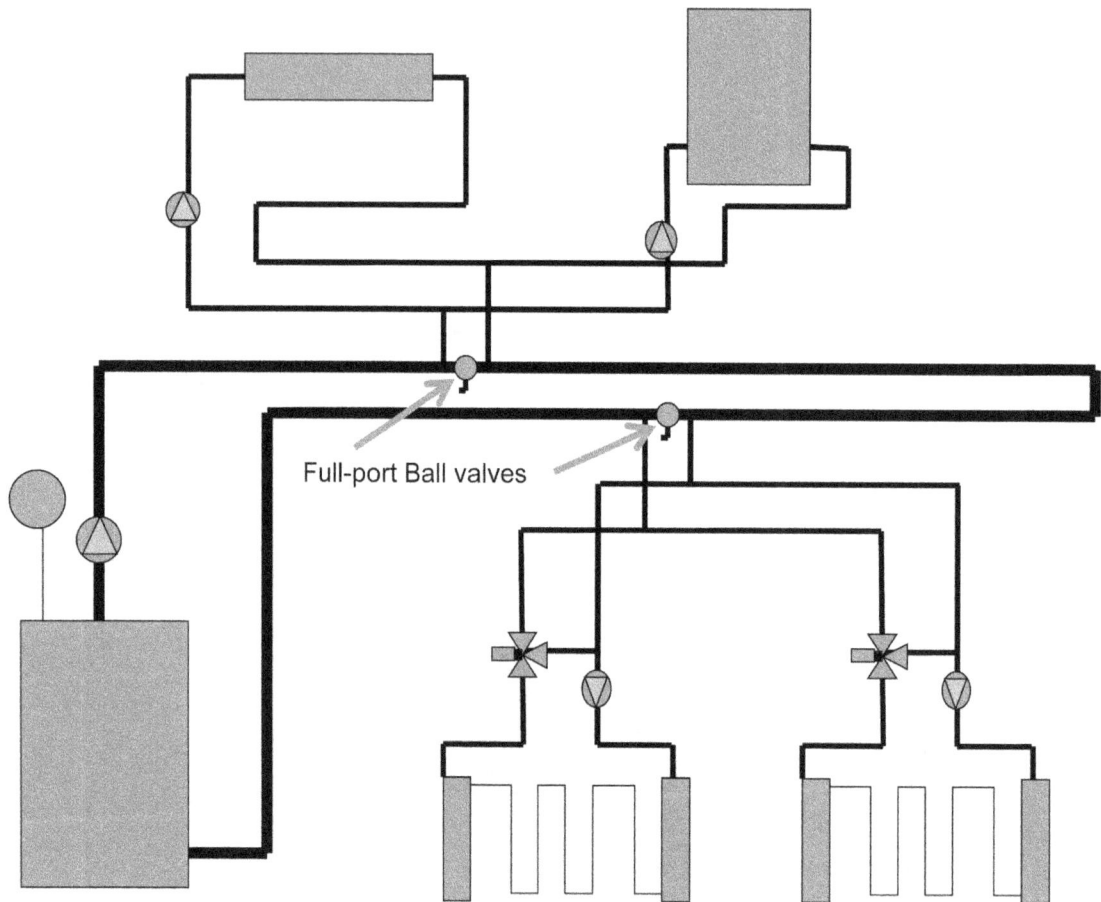

Full-port Ball valves

Without this valve, it's going to be very tough to purge air through those secondary circuits. But a full-port ball valve (which you'll use only on start-up) will divert the purge water from the primary circuit and into the secondary circuits and that can save you lots of time.

Once you're through purging, the ball valves stay open, and that's the reason why I recommend a *full-port* ball valve. You don't want to have anything in the common piping that can create a pressure drop because that pressure drop might encourage water to flow into the secondary circuit when the secondary circulator is off. That can cause overheating in the secondary. Full-port ball valves are as wide open as the pipe when you leave them open.

And while we're on the subject, I want to give a shout-out to our friends at Webstone. They make a lot of interesting valves, designed to solve the real-world problems that installers face. Here's one of those valves.

Webstone calls this a Primary-Secondary, Loop-Purge tee. It's everything you need to go from a primary- to a secondary circuit, and be able to get the air out of the secondary very quickly.

It's often difficult to get a classic, diverter-tee system started because air hides in the radiators, which are, in a way, secondary circuits, even though they don't have secondary circulators. When you purge the initial air from the pipes, the water will flow quickly around the main and back to your purge hose. You'll think you're done, but there's still a lot of air hiding in those side branches that lead to and from the radiators.

We have a lot of diverter-tee systems on Long Island and there once were hundreds of small, family-owned oil companies (such as the one Mr. Hicks owned) and those companies made lots of free service calls. That's the way it was back then. If you bought oil, you got a free service contract. The technicians knew they could spend a lot of time getting the air out of a diverter-tee system once they drained it to work on the system. They developed a few tricks to help themselves. It went like this:

1. If at all possible, pump away from the compression tank because that puts the circulator's full differential pressure on the air bubbles that are hiding in the radiators, making them smaller and easier to move.

2. If that's not possible (and most of the time it wasn't because the circulator was already installed on the return side of the boiler), raise the weight inside the flow-control valve by turning the knob that's on top of that device. This will lessen the pressure drop across the system and give the circulator a better shot at those bubbles.

3. Raise the system pressure to within a few pounds per square inch of blowing the relief valve. When you do this, Henry's Law kicks in. Remember, the higher the pressure, the more the air will want to dissolve into the water.

4. And following the same law (Henry's), lower the system temperature. Cooler water absorbs more air than hotter water.

5. Of course, once you get good circulation everywhere, you have to put the pressure and temperature back to where they should be.

Those oil guys specialized in getting the call done quickly and getting on to the next one, so give these tricks a try the next time you can't get the air out of that diverter-tee system. They work.

Dawn

Mark Eatherton, who is one of the sharpest heating contractors ever to pick up a wrench, told me about a trick that he's used to get rid of air bubbles. Here's Mark:

"This was taught to me by an old-timer many years ago, long before we had the under-standing that we have today, but it still works. On my own home, (one-pipe, diverter-tee system) I had problems with an entrained air bubble working its way around the loop. I tried everything, including power purging, tipping/trapping radiators, etc., but to no avail. I finally broke down and put in a micro-bubble reabsorber. The rogue bubble still kept bouncing around the system. The air separator just couldn't catch it. Oh, and I'm only moving 2 GPM through a 3/4" circuit, so it wasn't an excess-velocity issue.

"I finally remembered the soap trick, and injected an ounce of Dawn liquid detergent into the circulating main. The results were immediate. Within minutes, the system was virtually silent, and remains so to this day. The soap acts as a surfactant, breaking up surface tension, and allows the bubble to be broken up into smaller pieces, which are easier to capture and expel. I, and many others, have used this trick successfully on troublesome systems. A little soap is all you need, so don't overdo it. It also has the benefit of raising the pH of the system, which is always a good thing, except if you're using an aluminum boiler."

Mark told me that he carries a small Silver King hand pump and a bucket in his truck. He drains a gallon of water from the boiler into the bucket and then squirts in an ounce of Dawn. He mixes it up and pumps it back into the boiler. In a pinch, he'll drain the pressure off the boiler, pull the relief valve, and pour the mixture through that opening.

Just another one of those things to keep in your bag of tricks.

Air Control vs. Air Elimination

In 1970, when I started in the heating industry, there was a battle going on between the folks who believed in catching and controlling the air by holding it inside a plain-steel compression, and those who thought it was best to catch the air and just get rid of it, using automatic air vents and diaphragm-type compression tanks. As I told you, my father and I worked for the Bell & Gossett representative in New York at the time, so we had to subscribe to the Air Control school of thought because that's what B&G preached back then.

You can't be holier than the church.

The argument went like this: How can a closed hydronic system be truly closed if you keep venting air from it. When you lose air, cold water will take its place, and cold water contains more air (Henry's Law again). That air will come out of solution once the water gets hot. Automatic air vents (which we, as Air Control people, hated) would spit that air out and even more fresh, cold water would take its place. We saw it as an endless cycle.

The Air Elimination folks disagreed. They said it was better to vent the air and replace it with water, and even though the water contained more air, it was far less air than the amount you just vented, so it was a case of diminishing return. With each automatic venting cycle (they *adored* automatic air vents), less and less new water would enter, and before long, you'd have a truly closed system.

Put *that* in your pipe and pump it.

We fought the good fight, and so did they. We each had products to sell, and the older I get, the more I realize that that's what the fight was really all about. Ultimately, I'd say the Air Elimination folks won the war, and you'll probably agree with me. Just look at how many diaphragm-type tanks there are in the field these days, compared to how many plain-steel compression tanks you'll find in basements.

But a bit of background, and a story about another guy who taught me a lot, even though I never met him. His name was Ed Tidd, and he worked for Bell & Gossett when I was still in grade school. He had a hand in developing their line of what they called Airtrol equipment. (Airtrol was a shortening of the words, Air and Control.)

I first came across Mr. Tidd on a summer's afternoon when I was working in the rep's Long Island office. I was a customer-service guy and the phones weren't ringing much that day, so I took a walk out to our warehouse and found a carton of old books up on a high shelf. One of those books was a brown binder that had the title, *TIDD-BITS – Factual Reports of Hydronic Trouble Installations, with Tips for Their Prevention*. There was a drawing of a rabbit at rest near the top of the cover. The caption read, "Always on the alert for news of trouble jobs." Near the bottom of the cover, the same rabbit is leaping, and that caption reads, "He has spotted one and is on his way."

I was in love.

Inside this three-ring binder were hundreds of green pages. They broke down into two- or three-page reports, each about a problem job, and how the folks at B&G (mostly Ed Tidd himself, writing in the corporate "we") had solved the problem. Most of the problems had to do with air, and not surprisingly, Airtrol equipment played a huge rule in the solving of those problems. It was marketing at its best.

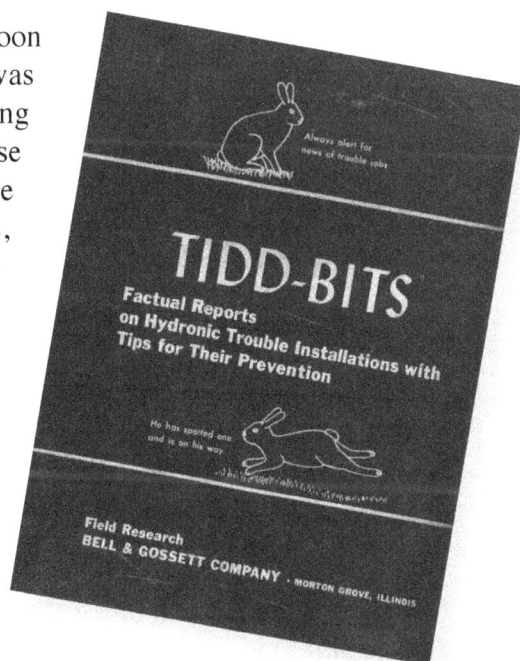

Ed Tidd would mail these reports to the reps as he wrote them during the early '60s, and the reps saved them in the binders that he also provided. It didn't take long to build a wonderful library of true stories, and what I loved most about these reports was the way Ed Tidd wrote. He spoke directly to me. He wove stories and used terms of simple English rather than engineering formulas. He caught me up in the drama of being on the job, and he made me feel the urgency of finding the cause of the problem because the customer was waiting and it was cold outside. He wrote to only one reader (me), and not to groups of people (how many people are helping you read this right now?). Ed Tidd just talked on paper. It was as though he was right there with me.

I wanted to learn how to do what he did, and that was the start of my career as a writer. If not for Ed Tidd and TIDD-BITS, I probably never would have written a magazine article, let alone a bunch of books.

Anyway, I talked my boss into letting me write a newsletter for local contractors, in which I would tell the stories of what we were seeing in the field (using the corporate "we"). I wanted to be like Mr. Tidd.

My boss said okay and we called the newsletter *The Problem Solver*. We didn't have computers or a mailing list, so I went to the public libraries all around New York, New Jersey, and Long Island where they had Yellow Pages from all the counties and boroughs. I hand-wrote more than 5,000 names and addresses of heating contractors from those Yellow Pages and that's how we got started.

The Problem Solver was popular, and in 1989, with a wife and four daughters, who were all going to be in college together in 2000 (and they were), I decided to leave the rep and see if I could start a business and make a living as a writer.

Thanks to you, Dear Reader, things have worked out okay.

Ed Tidd believed in the plain-steel compression tank, which I have to admit, still has a lot going for it. It has no moving parts. It hangs from the ceiling. It lasts for decades. Its biggest problem is that it will lose its air cushion to gravity circulation from the system, but Ed Tidd had that Airtrol Tank Fitting to offer, and that solved the waterlogging problem.

I know this is true because we moved into the house we're still in back in the summer of 1977. It had (and still has) the compression tank that the builder hung from the ceiling in 1950 when the house was brand-new. As I told you, our house used to have copper pipes in the concrete slab. They leaked after 20 years, so the guy before me abandoned the radiant and had all that baseboard installed.

Our first winter in the house was wonderful, but I could hear the air bubbles rattling through the baseboard loop. I solved the problem with an Airtrol Tank Fitting. It's still there all these years later because devices with few or no moving parts last a very long time.

Oh, and my tank doesn't waterlog.

I have seen problems with plain-steel tanks, but most of these problems are human-related. For instance, most of the plain-steel tanks that I've seen on commercial jobs have gauge glasses. The glass is there so someone can see the water level within the tank, but if you check the literature from the gauge-glass manufacturer, you'll notice that they tell you to keep the valves on the gauge glass closed unless you're checking the water level. That's because the valves on a gauge glass have stem packing, and that upper gauge-glass valve is in the air portion of the tank, not in the water portion. The valve's stem packing dries out, and air leaks right through it. As you lose the air cushion, the fill valve replaces the water, and that water goes into the tank because the rest of the system is already filled. The water level in the tank rises, and when it gets to the upper gauge-glass valve, it wets the stem packing and the valve stops leaking.

Standing on the ground, all you see is a waterlogged tank.

So you scratch your head, drain the tank and then repeat the process.

My advice is to get a ladder and climb up there. Close the two valves and use a hammer to break the glass. Problem solved. If someone replaces the glass, reapply the remedy.

Another problem I've seen, and this one is always amusing, is when someone drains a hydronic system without opening the vent at the bottom of the Airtrol fitting to allow air to enter the tank. As the system drains, the level in the tank begins to fall because there's this thing called gravity.

But there's also this thing called Boyles Law, which tells us that when you have a trapped gas (such as the air in the compression tank), and you allow that gas to expand (which is what it does as the water level drops within the tank), the gas's pressure is also going to fall.

When the pressure inside the tank gets to a point where it's less than the atmospheric pressure on the outside of the tank, the atmosphere will win. That's because high pressure goes to low pressure. Always.

You'll notice this happened because the tank, hanging up there from the ceiling, suddenly, and quite violently, caves in like a crushed beer can.

Whoops!

One more problem with plain-steel tanks. You can have as many of them as you'd like, but they all have to hook up to a single point of no pressure change. Remember we talked about this when we were talking about circulators? The point where the compression tank connects to the system is the one spot in the system where the circulator can't change the pressure. The circulator uses this as a sort of hydronic bookmark. It moves its differential pressure around in response to the location of the point of no pressure change.

But if you present a circulator with two or more tanks, each hooked up in a different spot in the system, the circulator will get confused and establish a point of no pressure change somewhere *between* the two tanks. That means there will now be a difference in pressure between the two tanks, so water will flow from one tank into the other. One of those tanks is going to get waterlogged. Guaranteed.

And you don't even need two or more actual tanks for this to happen. A friend of mine told me about a job where they were working with an old gravity system. He had added a circulator, and while he was venting air, he missed this one air-bound radiator that was under the sink. That radiator had enough internal capacity to act as a second point of no pressure change for the circulator. It caused his true compression tank to lose its air cushion and waterlog.

So if you have multiple tanks on that classic hydronic system, follow these rules:

- Come off your air separator in that Air Control system and pitch upward toward the tanks.

- The vertical line between the air separator and the tanks should never be smaller than 3/4".

- The horizontal run to the tank manifold should be ¾" if it's up to 7 feet long, 1" if it's up to 20 feet long, 1-1/4" if it's up to 40 feet long, and 1-1/2" if it's up to 100 feet long.

- As for the tank manifold, use 1" if you have two tanks, 1-1-4" if you have three or four tanks, and 1-1/2" if you have five or more tanks.

There's a bunch of stuff to consider when you're using multiple, plain-steel tanks, and that's probably why we see so many diaphragm tanks out there in the real world.

Let's take a look at that whole school of thought.

Air elimination

First of all, diaphragm tanks are smaller than plain-steel compression tanks, which is nice because boiler rooms are usually stingy with space. This is probably the main reason why they're so popular in modern boiler rooms.

The concept of these tanks is to separate the air from the system water with a rubber membrane, and to precharge the air side of the membrane to 12 psi at the factory. But as I've said, always check that pressure with a tire gauge before you mount the tank on the system. Just like the tires on your car, the tank can lose its pressure over time. In fact, according to Amtrol, a major manufacturer of diaphragm-type tanks (they call them Extrols), air will move across the rubber diaphragm by diffusion at the rate of about 1 psi per year.

Did you know that?

Diffusion is what happens when there's more of one sort of molecule, say nitrogen or oxygen, on one side of a semi-permeable membrane than there is on the other side. Since the rubber diaphragm is a semi-permeable membrane, the air just says "Excuse me!" to the rubber molecules and sneaks through to the side that has the water. From there, an automatic air vent will spit the air from the system. Fresh water will enter to take the exiting air's place and that's usually how diaphragm-type tanks fail. If you don't replace the air, the expanding water will tear the diaphragm off its anchorage and the boiler relief valve will pop.

A similar thing happens with party balloons. If you blow up a regular, latex balloon, it will stay full for a while, but then slowly lose its air. A regular balloon is a semi-permeable membrane. The air moves right though it as the balloon tries to contract.

Those shiny Mylar balloons hold helium a lot longer because Mylar is less permeable. Sure, a Mylar balloon will eventually get saggy, but it takes longer to do so than a latex balloon. Ever notice how the material they use for potato chip bags looks a lot like Mylar? The less permeable the membrane, the fresher the chips will be.

And by the way, this is the same issue that plagued the radiant-heating industry 20 or so years ago. It's why the manufacturers of plastic- and rubber radiant tubing put oxygen-diffusion barriers on, or inside, the tubing. Without the barrier, air will move through the tubing and attack the steel and iron parts of the radiant system.

Diaphragm-tank manufacturers give you a Schrader fitting on their tanks so you can check, and if necessary, pump up the tank to the proper pressure for the job. Again, that pressure must be whatever the system fill pressure is (it's so important). If you're filling the system to 18 psi because the building is three-stories tall, you'll need to pump up the tank to 18 psi because the factory only puts 12 psi in there. Do this while the tank is disconnected from the system because if there's water pressure on the other side of the diaphragm, it will squeeze the air and make you think there's more pressure inside the tank than there actually is.

I once visited Amtrol in West Warwick, Rhode Island and asked about that 12-psi precharge. I was thinking about contractor friends who work in Breckenridge, Colorado and places such as that, where the air pressure is a lot lower than it is at the Amtrol factory. That Rhode Island 12 psi isn't going to be 12 psi when it gets up in the Rockies.

They smiled and said that's what the Schrader valve is for.

So use it.

The deep-drawn tanks that Amtrol makes nowadays showed up in the 1960s. Before that, they had a neat history. To prove that the concept worked, they built the tank pictured below.

If it looks strangely familiar that's because they made this first tank from two streetlight blanks and a sheet of rubber. The rubber separated the air from the water, and the Schrader valve was there for the air pump.

It worked.

Their first commercially available version of the tank had a flying-saucer shape. They trapped a formed-rubber diaphragm between the two halves, and then crimped the edges to lock the tank together. This tank had welded legs so

the Dead Men could screw it to the floor joists in place of the plain steel tank they were replacing. The tank didn't have to hang from the ceiling, but old habits die hard, so why not accommodate what your customers expect?

During the '60s, they came up with this tank in the photo on the right.

They use a hoop-ring method to attach the diaphragm inside this one. I watched them make these and the machine that does the deed is a true beast. You feed a flat disc of steel into its mouth and, using hundreds of tons of pressure, it turns the flat disc into a steel dome. The strength of the steel doubles as a result of cold working, and that results in a lighter, stronger tank. Watch it and wince.

One of the Amtrol guys told me a fun story about their first cylindrical Extrol tanks. They were using an overstock of baby-carriage rims to manufacturer their tanks in those days. When the competition arrived and came up with their version of the product, Amtrol bought a few and took them apart to see what was what, as any good competitor would. They found small holes drilled into the competitor's steel hoops. Those guys were mimicking what they thought was a necessary feature, but it was just the spoke holes for the baby carriage rims.

Engineers. You gotta love 'em.

The early Extrol tanks didn't have an air valve. To minimize potential leak paths and speed production, Amtrol charged those tanks with a measured chunk of dry ice prior to assembly. At room temperature, the dry ice evaporated and produced a 12-psi charge of carbon dioxide.

But then there are places such as Breckenridge, CO, and that's why you now see Schrader valves on these tanks.

As for sizing, tank companies, such as Amtrol, make it easy. Just go to their Web site and use the online calculator. You'll need to know the system volume in gallons, the difference in pressure between the fill valve and the relief valve, and the average water temperature. If you're using a fluid other than water (say, antifreeze), you'll have to correct for that, but it's easy to do.

Following the Air Elimination (not Air Control) philosophy means that you're going to spit out all the air from an air separator, which has an automatic air vent of some sort. There will probably also be automatic air vents at the high points of the system.

Let's take a look at those air separators because you'll see all sorts of them on classic hydronic systems.

Separating the Air

Back in the day, when boilers had lots of internal space, the Air Control camp introduced the boiler dip tube. Here's one (on the right) that fit into the top of a boiler.

It's a simple device with no moving parts. The boiler is the wide space in the road, and the water's velocity will slow as it enters the bottom of the boiler. When water moves slower than six inches per second, free air tends to come out of solution and rise to the top of the flowing water. When it enters a boiler, particularly an older boiler, it will rise to the top and move by its own buoyancy through that opening in the Airtrol Boiler Fitting, and up into the plain-steel compression tank. The water that flows out to the system has to take a detour to the bottom of that dip tube, and air is too buoyant to follow. Some older boilers had side-outlets, so the Air Control folks came up with a device that would work the same way, but from a different angle.

As boilers got smaller and more efficient, they ran out of room for this sort of dip tube. Using the same idea, though, some boiler manufacturers cast a dip tube of sorts into their end boiler section. You'll see this right next to the supply tapping at the top of the boiler. Always put an automatic air vent in that hole, or connect a pipe from it to your plain-steel compression tank. Don't put a plug in that tapping because air will gather under it (after all, it *is* an air separator) and that can cause that part of the boiler's section to overheat. The iron is hot and if water's not flowing over the metal to transfer the heat, the metal will suffer.

The next thing we see is an external version of that internal dip tube. They just put the boiler dip tube into an external tank and it worked because the water would slow as it entered that wide space in the road. Air would go to the top of the tank as water went lower to enter the dip tube. Here's a sketch of the setup. This is from an issue of Tidd-Bits.

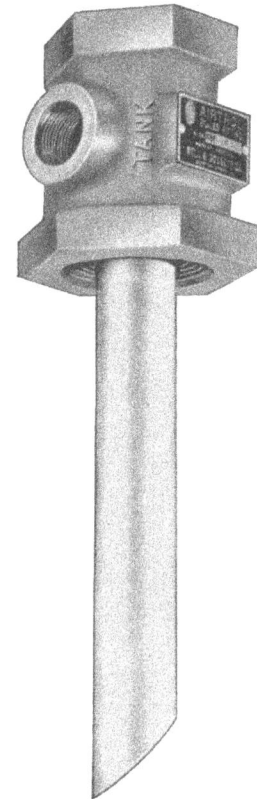

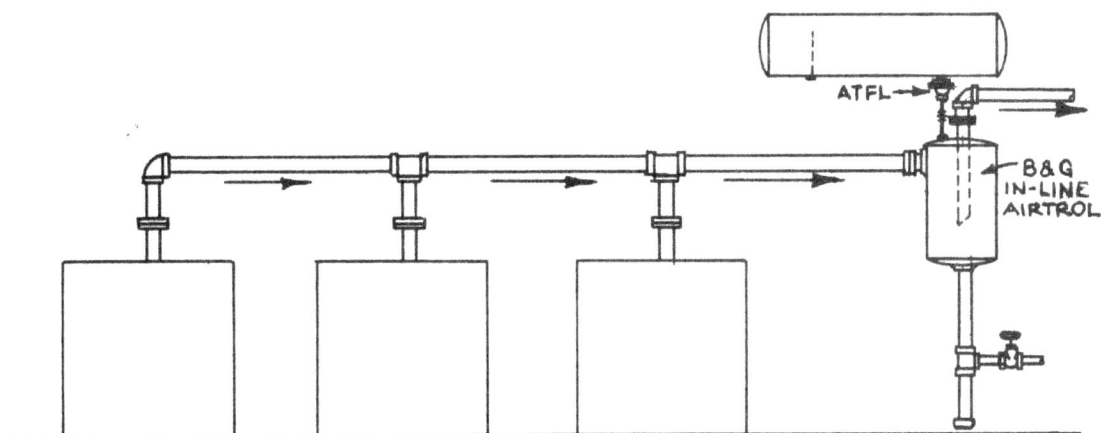

Air scoops show up around the same time, and some had a tapping for a plain-steel tank, but with Amtrol's introduction of the diaphragm tank, the air-scoop manufacturers dropped the tank tapping in favor of a smaller tapping that would hold an automatic air vent. Air Elimination guys at their best.

Note the tapping for the tank on this early air scoop from Taco.

And this later one from Amtrol, without the tapping for the tank.

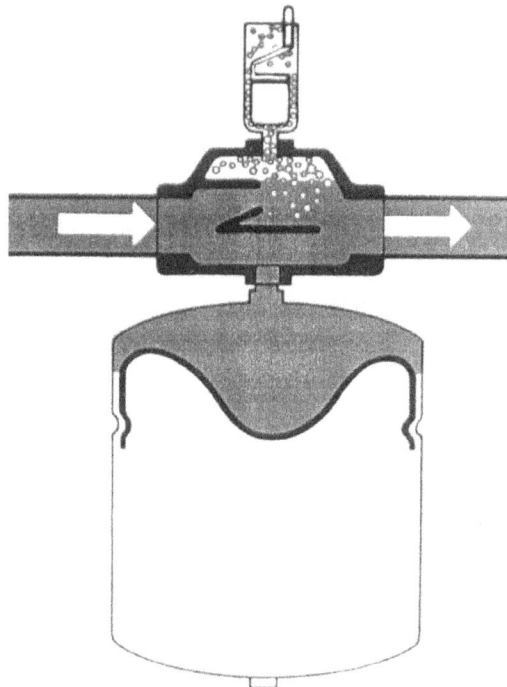

Chapter Ten: Classic Air Problems

Air scoops work well because all the water has to pass through them again and again, but the manufacturers of these scoops do like to see at least 18 inches of straight, horizontal, full-size, unobstructed pipe before the scoop. That's so the water can calm down and flow in a laminar way. The air can then rise to the top of the moving water and be scooped toward the automatic air vent.

In the field, most scoops don't have that manufacturer-specified piping. Most connect to the vertical riser with a street elbow.

It's a regular reality show out there, my friend.

And that brings me back to microbubbles; those microscopic bubbles that make the water appear milky. Those tiny bubbles like to gather together and form bigger bubbles, and that's where these devices called microbubble reabsorbers come in.

Remember that Brooklyn plumbing supply house I mentioned earlier, the place where I first saw the Spirovent? That was during the summer of 1992, and the salesman was a delightful guy named Danny King. Sadly, he's no longer with us, but he was one of the most enthusiastic guys I've ever met. He could get excited over ice cubes, so you can imagine how he got with this new type of air separator that could catch tiny bubbles.

Spirovent is a product of Spirothem, a family business with its roots in the Netherlands. I visited there once and watched them make these things. They have a true passion for what they do and they love to tell about it. That was good because I love to listen.

On that long-ago day in Brooklyn, Danny was demonstrating the product on this see-through plastic rig that the company had put together. It was a loop of clear piping with a circulator, a see-through Spirovent, and a bicycle pump. Danny had drawn a crowd and he encouraged this burly Brooklyn heating contractor to pump air into the loop. The guy did, and the Spirovent just spit it out. The burly contractor got this look on his face that was pure bliss. I'm telling you, he turned into a boy. Another contractor would want to try the bicycle pump, and the burly guy would give him a dirty look. *My* toy.

I watched this for a while and realized that these folks knew how to get through to classic contractors.

You have to *prove* it.

They did.

So that product became popular, and other companies followed with devices that worked in a similar way. No special approach piping needed for any of these. The microbubbles are just going to get stuck on the nest of thin wires, and then rise by their buoyancy to the air vent.

It's Air Elimination on steroids.

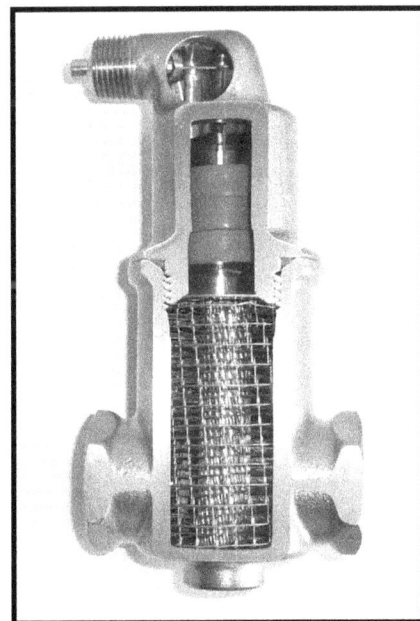

Spirovent

But because these work on Henry's Law, as do all air separators, I wondered whether they would work near the bottom of tall buildings, where the static weight of the water would make it more difficult to remove the air from the water. I asked a friend at Spirotherm and he told me this:

"We have job results where the customers told us how well the Spirovents performed in buildings (residence halls, etc.) up to 15 stories. In fact, I had to write a letter to a building superintendent in New York City several years ago to stand behind a 12-story, multifamily installation. He actually wrote back to let me know how happy the tenants were. The tallest building I know of with a Spirovent is the Imperial Tower in British Columbia. I believe that one is 30 stories. The unit is in the basement, although I think they also have some Spirotops (designed for the very quick removal of free air) in the penthouse. They were having some nasty air problems with their original air separator in the boiler room, and they had standard automatic air vents at the top of the system. The owner also came back to us on this one. Problems solved."

Real world. Real people.

Honeywell came out with a unit they call the Power Vent. It looks like this.

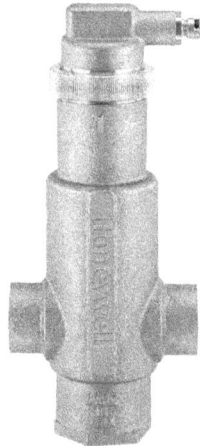

Bell & Gossett calls their version the Enhanced Air Separator and it looks like this.

Taco goes in a different direction with their Vortech.

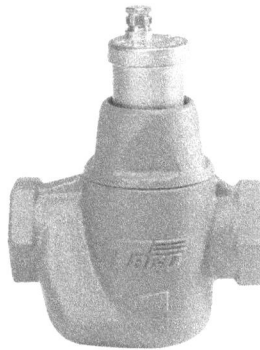

Water enters and leaves the Vortech on a tangent, so it spins as it moves through. That causes the lighter air to move to the center of the flow. The unit has what Taco calls a Bubble Breaker cartridge, which (you guessed it) breaks bubbles, and from there, the water has to duck under a ledge to get out, leaving the air alone to escape from the built-in vent. I wonder if the water says, "Weee!" as it goes through.

Many larger, classic-hydronic systems use a device that Bell & Gossett calls the Rolairtrol. These also work with tangential flow. The water spins, causing the air to move to the center, where it rises to a plain-steel compression tank, or leaves through a high-capacity air vent.

Note how the water has to squeeze past that round, horizontal baffle that extends nearly to the edge of the sidewall. Only the water with the least amount of air goes through.

Other large air separators use Pall Rings to catch the air. Here's an example of one.

Pall rings

Pall Rings catch tiny bubbles, and they have an interesting history that starts in Germany. I wrote to the company and this is what they shared with me:

"In 1891, Fritz Raschig, an enthusiastic chemist, decided to establish his own business. He was granted a license to build a chemical factory by the city of Ludwigshafen. Apart from the main business of tar distillation, he carried out intensive research to find a method of producing highly purified tar components such as anthracene, carbolic acid, and benzene. His efforts were eventually rewarded by the development of the revolutionary "Raschig Ring" - a milestone in process engineering. This led to Raschig producing Raschig Rings in ceramic and metal. These were pieces of tube with equal length and diameter.

"A better gas flow was required, and BASF Aktiengesellschaft, the large chemical manufacturer in Germany, took the Raschig Ring in the early 1950s and cut holes (windows) in the cylindrical portion to increase gas flow. This reduced the surface area so they left the metal punched out of the window area as a finger. For some reason they called this a Pall Ring. Perhaps Herr Pall was the engineer? In the late '50s, plastics were included as a molded Pall ring."

So there you have it. The company that makes them doesn't know why we call them Pall Rings and neither do I, but they do a nice job of catching tiny bubbles in a classic-hydronic system because they present the air bubbles with plenty of surface area on which to stick.

Air vents

Manual air vents, the sort that you open with a key or a screwdriver, have been around for as long as hydronic heating has been around. They're good for getting rid of air at high points, and they're a full-fledged member of the School of Air Control, where a truly closed system means *no* automatic air vents allowed.

Don't try to vent any radiator or system high point when the circulator is running. It's very difficult for the air bubbles to grab onto the air vent and yank themselves out of the system when they're going with the flow at the rate of about four feet per second.

Use your imagination.

If you have old, key-type air vents, and you're missing the key, check out the Internet sites that sell the keys for winding old clocks. These usually work on old air vents, and most are very well made.

And once again, remember that the radiator with the toughest air problem will *always* be the radiator that's behind the heaviest piece of furniture. Or the fish tank.

It's true.

Keep in mind that while venting, if you don't get any air, it ain't an air problem. It's a balance problem. Stop venting because if you continue to vent you will receive the reward of a temporarily warm radiator. You get that reward because you have done something dumb.

Stop venting.

Automatic air vents come in all sorts of shapes, sizes, and price ranges, and they often prove the case that you get what you pay for. Some of them will clog easily and leak. Many of them have caps that you can screw down to prevent the water from leaking, but this will also prevent the air from venting, so go figure.

Some automatic air vent manufacturers offer a kit, which you can use to run a small tube from the vent to a location where the vent can drain, should it decide to leak. This can save a drop ceiling from dropping. It can also solve problems with callbacks, which reminds me of another story.

I was driving to a problem job with a New Jersey contractor many years ago when he slowed by this old, brick apartment building that had ivy growing up its walls. He pointed at a top-floor window and I noticed what looked like fog coming out of the ivy, just below the window.

"See that?" he said. I nodded. "That place has a one-pipe, diverter-tee system, and I kept having to go back to get that one radiator hot."

"Did you get any air when you bled the radiator?" I asked.

"Never," he said, "but the radiator would get hot."

"That ain't an air problem," I said. "That's a balance problem."

"Well it sure *was* an air problem," he said. "You know how I know? I solved it with an air vent."

"How did you do that?" I asked.

"You ever see those kits they sell with the automatic air vents? Those little tubes that you can run to a drain?"

"Sure."

"Well, I connected one of those to that radiator's automatic air vent, drilled a hole through that brick wall over there. Yeah, right there behind the ivy. Then I ran the tube thorough the hole and into the ivy. I caulked it up real good and that's the last time I had to go back."

"Looks like it's good for the ivy," I said.

"It's *all* good," he said.

You can't make this stuff up.

A final thought about air

In Europe, nearly everyone has hydronic heat, and for years, those who make the laws over there have mandated continuous circulation with outdoor-air-temperature reset, which is sort of like cruise-control for your heating system. Nowadays, they're also insisting on variable-speed pumping, which keeps the water moving all the time, but at variable flow rates, depending on the heating needs at any moment.

We have no such laws in the U.S., and the choice of variable-speed pumps, continuous circulation, or start-and-stop is up to the installer and the customer. How much do you want to save? And how much do you want to spend?

Most classic-hydronic systems run with a thermostat that starts and stops a circulator, which runs at one set speed. When the water stops moving, any air that's in the water has a good chance of coming out of solution and working its way up to the top of the system, where it can get trapped in a radiator. This is especially true if the system is out of balance.

I think this is why air is a much bigger deal in American hydronic systems than it is in European hydronic systems. I don't see any new American laws in the works that will ever change this.

That's all the more reason to use good air control, or good air elimination. It's your choice.

CHAPTER ELEVEN

Classic Radiant

The Lovely Marianne and I grew up in Hicksville, Long Island. And please don't laugh. Hicksville had about 42,000 people living there at the time. There's not much "hickey" about the place.

I spent a good part of Junior- and Senior High School sharing a classroom with Billy Joel, which, with the exception of the day I shook hands with Elvis, is my only brush with Rock-N-Roll royalty.

Here's a picture of Billy from our 1967 yearbook at the bottom right.

They took that picture at a high-school dance. When the band took a break, Billy approached my high-school honey, who was from a neighboring school. He tried to smooth-talk her, but she just gave him a sweet smile, told him she liked the way he played the keyboard, and then explained that she was "with Danny."

He gave me a tip of his cap and moved on.

She and I broke up right after high school, and I imagine she sits somewhere nowadays, smacking herself in the forehead.

Anyway, Billy and I sat next to each other in Biology class when we were about 15 years old. Under his breath, he was singing a brand-new Herman's Hermits song called Mrs. Brown You've Got a Lovely Daughter, and he was tapping his desk with the eraser ends of two yellow pencils. He smiled at me and proclaimed that some day he was going to be more

famous than Herman. I smiled and nodded because even though we were just a couple of sub-urban kids, anything seemed possible in those days.

Billy Joel lived in Levittown, which is right next to Hicksville. Hicksville is right next to Bethpage, which is where we live today. Down the spine of Long Island, there is a smear of towns that tumble and fold over each other in a sprawl of little homes that were all built at the same time in the days following World War II.

Our fathers returned from the War and moved their young families from New York City to the "country," where they bought or rented these inexpensive homes, and set out to live the American Dream. They had earned the right.

About 17,000 of these cookie-cutter houses were in Levittown, and these were the first radiantly heated tract homes in America. Thousands more of these little dwellings wound up in Hicksville and Bethpage and another Levittown in Pennsylvania, and radiant kept many of us cozy while we were growing up.

Here's a photo of what Levittown looked like when they first opened it.

When people returned from work, they sometimes couldn't figure out which house was theirs. Nowadays, no two Levitt houses are the same. People built onto them and added baseboard heat to the upstairs, and Levittown is now a pricey neighborhood. Not long ago, the Smithsonian tried to find an original Levitt house to transport to Washington D.C. but there are no longer any originals.

In 2003, I received a letter from Carol Blum, a fellow Long Islander. Her father, Irwin "Jal" Jalonack, was instrumental in the development of radiant heating in America. Jal was the guy who made the decision to use hydronic radiant heating in Levittown. This was in 1946, and here's the story that Mrs. Blum told me.

"My father, Irwin "Jal" Jalonack began his work life as a plumber. He apprenticed to his father in Syracuse, New York during the early 1920s. Later, he became an HVAC engineer and, as William Levitt's Executive Vice President, made the decision to use oil-fired, hot-water radiant systems in the houses of Levittown.

"During the 1930s, my father patented something called a "Jal fitting." I think it had to do with heating pipes. Have you ever heard of it, or seen one? If you have, I would be interested in tracking one down. I realize you tend to look at work that is a bit older, but Levittown is now over 50 years old, so I thought some of its workings may have piqued your interest. By the way, my brother and I also grew up on Long Island - in Old Westbury and Roslyn around the same time you were growing up."

I wrote back and told her that I had a "Jal" fitting on my desk. It's nothing more than a ¾" X 1/8" bushing that fit into a tee on the return side of the system. Here 'tis.

Whatever water could fit through that 1/8" hole would flow from the radiant system into the boiler and be heated to 180 degrees. The rest of the flow would bypass the boiler and join the heated trickle on the supply side to head back out to the floor. It was marvelously simple temperature control, and it was perfect for the times. We live in a house that once had one of her father's heating systems. Here's a rough sketch of how the system works.

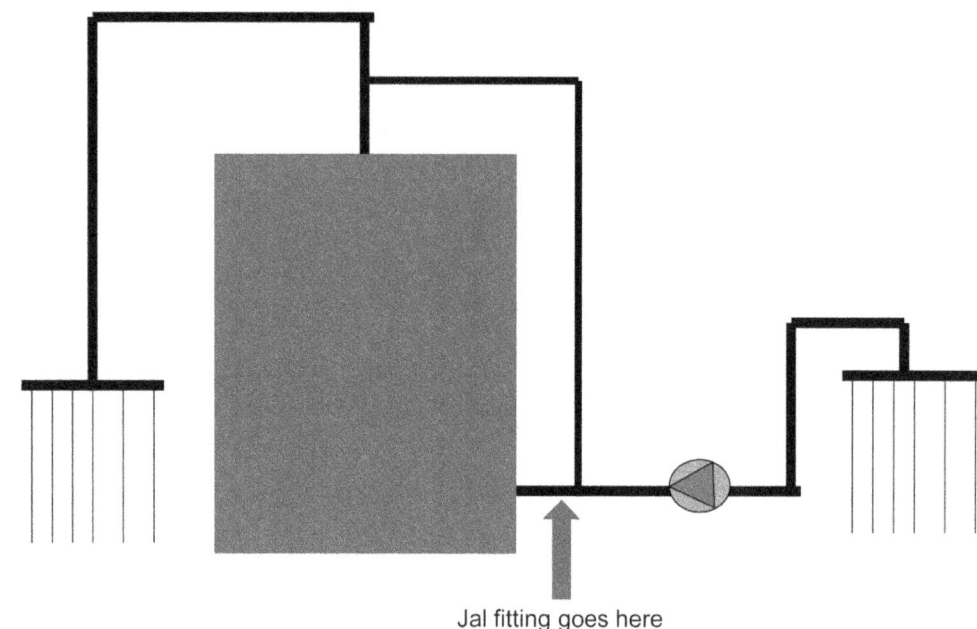

Jal fitting goes here

If you set the boiler to maintain 180-degree temperature (these boilers had tankless coils for domestic hot water), the supply temperature to the floor would be 140 degrees. That seems hot by today's standards, but keep in mind that people liked to put rugs on their floors back when these houses were new. Levitt floors emitted 50 Btuh per square foot. Those rugs were thick!

"Both of my parents are now deceased, so I can't get the whole story as to why he went with radiant heat. I do know that, at the time, my father was considered one of the experts in this country in the use of radiant heating. However, I do know a little that I can tell you."

I know that he chose radiant because it was, as I said, cheap. All of the Levitt houses are slab-on-grade construction and mass produced. They were building these houses like Liberty ships. They had to keep costs low. There were people whose only job was to install door knobs. Radiant was cheap because they laid the soft copper tubing right on the ground and poured concrete over it. Here's a photo of that.

There's no vapor barrier under the slab, and no insulation around the perimeter of the slab or under the tubing. If a leak developed (and they did), water flowed into the ground and not up onto the floor. Homeowners had no idea the leak was there until things got bad.

See in the photo how they attached the copper tubing to those wooden studs? That was to keep the copper in place during the pour, but once those studs rotted, the tubes were in a perfect position to crack when the slab settled.

Back to Mrs. Blum:

"Of course, the idea was to build an inexpensive house that could be run economically. Clearly, building on a slab was the way to go. My father felt that forced air was not good. He also preferred oil to gas. He always said that, were it cheap enough, he would have preferred to use electric heat, however. In fact, he had many discussions with the Long Island Lighting Company about this possibility of electric heat for Levittown. But luckily, as it turned out for the Levittown residents, LILCO would not come up with a cheap enough electric rate.

"My father made the decision to go with oil heat and then went looking for an oil-fired boiler that was small enough to fit in the kitchen, next to the other appliances. Since Levitt owned many of the houses, and rented them out, my father had to find an oil company that could handle supplying oil heat to the 17,000 new homes. He chose a small two-truck company that he thought would be able to grow and do a good job. You have most likely heard of them. They are Meenan Oil."

Just so you know, Meenan went on to become an enormous fuel-oil company and they continue to service Levittown.

"My father said that some of the problems with the leaks in the Levittown radiant systems had to do with the way the workmen had installed the copper tubing. If it had too much play in it, it broke faster. In 1965, his solution for the serious leaks in the copper tubing was to suggest sealing it off and installing baseboard convectors. I suppose that today they use plastic or some similar type of tubing, rather than metal."

As I drive through Levittown during the winter, I can tell which homes still have working radiant systems because there's no snow within several feet of the house. Oh, and flowers bloom very early in the spring. Like in late-February.

"My father was a hands-on kind of guy. Besides being VP, he was in charge of all purchases. Although he had an office at Levitt's main office in Manhasset, he had another larger one at Levitt's office in Roslyn. If you know where the railroad trestle crosses the Long Island Expressway between Albertson and Roslyn, just off the north service road, on the west side of the tracks, there was a building with a railroad siding. In those days, the road dipped as it went under the trestle. And of course, when it rained, the dip in the road always flooded. We all knew it as Lake Levitt. Pop lost one car (either a Nash or a '48 Hudson) to Lake Levitt."

"From his Roslyn office, he would set out to Levittown and ride around. The workmen never knew where or when he'd show up in his battered black Hudson to check on construction. As you probably know, the exact amount of supplies, in exactly the correct sizes, would be dropped in front of each plot for the builders.

"Pop said that the tradesmen, especially the plumbers, were not used to having things cut to exact lengths for them. If the plumbers took a longer pipe than they needed and cut it, they would later find that they were missing a longer pipe that they needed. They eventually got used to this method of construction, though. In fact, my father said that they would speed up as the day went on. So while they could do one to two houses in a morning, by afternoon they would usually get to three more. They all knew him as "Jal," and years later, when one of the

Meenan men came to our house to fix the burner, he looked at my father, yelled, 'Jal!' and they had a good time talking about the Levittown days.

Imagine installing five radiant heating systems in a day, and doing that every day of the week.

"In 1950, we moved from North Park (a small Levitt development off Willis Avenue) to a house my father built, located on Potters Lane in East Hills. The house, which was demolished to make way for the Long Island Expressway, was on a slab and had radiant heat in the floors. When he knew that the State was going to take that house, he built another one on Meadowbrook Lane in Old Westbury. It also had radiant heat, except that since it had a basement, the radiant heat was in the ceiling. There was also central air, but for that, he had his own AC well, and a duct system throughout the house. That house is still standing.

"After he left Levitt, my father built homes himself under the name of Jalco, and then with a partner as Jalfeld. He built in Commack and Deer Park. In those houses, he used oil-fired, hot-water heat with baseboard convectors. At some point during the late 1960s, he also worked with Andrew and Johnnie Levitt when they built under the name of LevittHouse, out in Stony Brook, Long Island. Andrew and Johnnie were Alfred Levitt's sons. Of course, Alfred was Bill's brother. I remember all of them, including Pop Levitt himself. And did you know that Levitt almost built homes in Israel in 1949? But that's another story, and it probably wouldn't have included much heating."

During the 1950s, some of the plumbers on Long Island began to specialize in the repair of the radiant heating systems. Here's an ad from one of those plumbers.

They gave homeowners a map of their radiant pipes and put lots of scary thoughts in the homeowners' heads. If the homeowner's fuel bill or water bill suddenly rose, he probably expected a leak in the radiant piping, so he would call a contractor such as J.D. Brower, who would show up and either listen for a leak with a stethoscope, or perhaps mop the floor and watch as it dried. That usually showed where the leak was because the floor would dry in a circle rather than a straight line over the leaking pipe.

But many of these people had rugs and you can't mop a rug or listen very well through a rug, so the contractors also kept one other very important tool in their bag of tricks.

The lazier the better. Cats will always find the warmest place to settle, and the old-timers told me that the cat hardly ever missed the leak.

By the way, this is where we get the term CAT scan.

(Whoa!)

The boilers went in the kitchen, either on the floor, under an appliance cover, or on a shelf. Here's the York Shipley model on the floor.

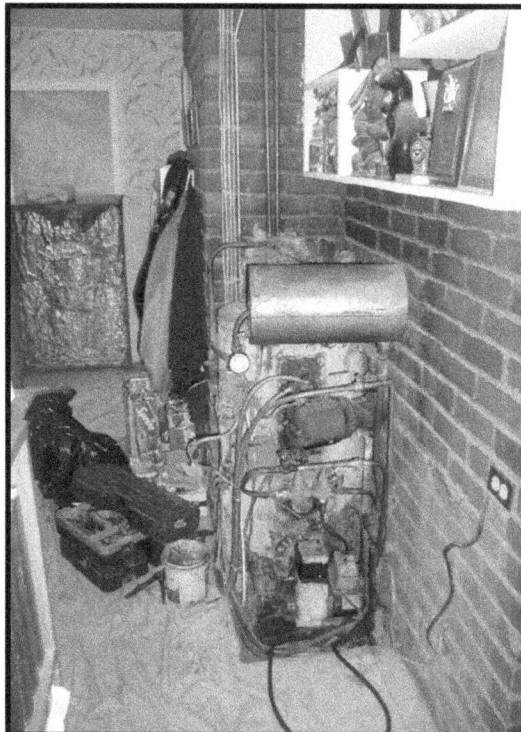

And here's one by General Electric, which fired from the top to the bottom.

They hired thousands of workers at a time when Americans were coming home from the war. None of those workers had to be highly skilled because the Levitt brothers were following the advice of Henry Ford, who had said, "In mass production, there are no fitters."

And that changed everything.

The Levitts set up a warehouse for supplies and appliances right there on the site. They built a complete woodworking shop, where 70 men cut lumber for 40 hours a day. That's all they did. They cut piece after standard piece because, although there were five different architectural styles for the Levitt houses, all the interiors were identical.

The Levitts set up a plumbing shop right on the site to prefabricate plumbing trees. They did the same for the radiant heating systems. I have the heating plans for these homes in my files. They first used soft copper, and when this became temporarily unavailable, they used a flexible steel tubing that General Motors made for them.

Oh, and they had a sand, gravel and cement plant on the site.

Why not, right?

The Levitts got the Federal Housing Authority to approve the use of plywood for their roofs and they used this for the first time ever in home construction. For the exterior siding, they used 73 large sheets of Colorbestos on each house. Johns-Manville developed this for

Levittown. The large sheets replaced 570 small asbestos shingles, which they had used on the first houses.

One day, Alfred Levitt was buying hamburgers at White Castle when he noticed the sliding Thermopane window in the aluminum frame that the kid was using to keep himself warm while the customers waited outside. That White Castle window is the reason why all the Levitt houses had sliding Thermopane windows, and that was another first.

No painter ever used a brush inside these new homes. They just walked through with spray guns and hit everything in sight with flat-white. And when the painting was done, other workers arrived to lay the black asphalt floor tiles. I know these tiles well because I've lived in these houses. When the tankless coil fouls, the service technician from the oil company raises the aquastat setting to 210 degrees so you can have a hot shower. This raises the water temperature flowing through the radiant coils to about 170 degrees. That's when the asbestos tiles slide around the kitchen like hockey pucks, and the pets jump up on the chairs.

Make sure you wear shoes.

The Lovely Marianne and I grew up in Hicksville, which touches Levittown, which rubs up against Bethpage. We played on warm floors and we all believed that anything was possible.

And it is.

CHAPTER TWELVE

The Radiator

This radiator, the one by the front door, was born in the mind of a young man, now dead, on a day in July in a small town in New England. He had grown up drawing and tinkering, and that had led naturally to this.

He spent most days dreaming and imagining, and today he was working on the radiator. It would be functional, of course. It had to be that, but if he did his job well, he knew he could also make it beautiful. He was, in a way, a sculptor, and even though he knew that most people don't notice the beauty in simple things, he still did his best because this was his art. This simple radiator.

The young man's design sprang from his mind and his hands and became a mold, and then one day in the foundry, other men pounded sand into his mold and then poured molten iron, as bright as the sun, and the young man was there to watch the birth.

It took hours for the section to cool, and when they shook the sand from it and cleaned it the man was there to see what had once been just an idea. And he thought it was beautiful.

They cast many more sections that week, and they joined them together with iron nipples that had threads running clockwise and counterclockwise. The men grunted in the heat as they did this work, and they didn't give much thought to the art. It was good, hard, American work and it was steady work, and that was what mattered. There were families to feed.

And there was this traveling man. He made his way up and down the east coast by train. He sat in sweaty railcars with other traveling men, smoking and swapping stories. He was garrulous and people liked him. And he was also like a dray horse in the way he traveled, as were all good traveling men in those days. He was as regular as clockwork. From city to city, it was always the same. He'd arrive at a place and take rooms in the same hotel as always. The trade press would announce his arrival in their weekly magazines. They would let people know that he would be in his rooms downtown, and that he would be receiving prospective customers between these hours.

He had brought literature and small samples of his product, and he would be showing a new radiator on this trip.

The buyers came and they liked the radiator. Surely the well-to-do, who could afford this new central heating, would feel the same, so they placed their orders and waited for delivery.

The radiators also traveled by train, and then by wagon to the supply houses where burly men grappled them into place so that the builders and the architects and the tradesmen could come and look. And these people liked what they saw. The radiator was beautiful!

A fitter worked with his apprentice one day in a fine house not far from downtown. Twenty of the new radiators would warm this house, and the fitter was pleased because he had gotten a fair price. He had figured sixty days labor for this job because that's how they did things then – three days labor per radiator for a fitter and his apprentice. This allowed enough time to cut the pipe by hand, and to do right by the many angles and turns in the piping that the water, traveling by gravity, required. It was a trade, and they were fitters, and it took time to do the job well. They were a prideful people, these fitters.

The family moved into their new home and they lived there for many years. There were children and grandchildren, wars and quieter times, sickness, health, good times and bad. They lived there through hot summers and many cold winters and the radiators served them well for all those years. Father would come home from business and that radiator, the one by the front door, would greet him before anyone else in the family. On frigid days, he would turn toward it as soon as he walked through the door. He'd take off his gloves and his wet hat and he'd place them on top of the radiator to dry, and then he would hold his palms toward the iron and absorb its warmth. He hardly noticed the radiator's delicate iron embroidery anymore.

Time passed and another family came to live in the house. It's like that in America. People live their lives and leave, but the old houses, with their old radiators, remain and wait patiently. The new family was different from the old family in some ways, and similar in others. They moved in and never really noticed the radiators. These were just fixtures to the new people, a part of the house. They were functional and no one really noticed them anymore, except when the heat came up in the fall "I smell the heat," the mother would say, and the father would smile. "Must be fall!"

It was a surprise now, the day the heat first came up. It wasn't that way with the family who first owned the radiator. Back then, Mother would have to make the fire each morning. She knew when the heat was coming. No surprises then, but this family had an oil burner. They were a modern family, and they spent little time thinking about the heating system, or noticing the radiators.

The old dog did, though. He noticed the radiators, especially this radiator here, the one by the front door. He slept on the throw rug in front of that radiator on most winter days, taking in the glorious warmth, waiting for the man to return from work.

The children grew and left and returned with their own children, and one day little Molly touched the radiator by the front door and burned herself. She let out a screech that had everyone running at the same time. Her mother scooped her up and comforted her, and then laughed, remembering. "I did that once," she said to Molly. She smiled at her father, and he smiled back. "But only *once*, sweetie. I did it only once. Be careful. The radiator is hot."

Molly looked through the shelter of her mother's arms and at the old radiator. "Bad!" she said, and her mother laughed again.

More years passed and each family left memories in the old house. The radiator, the one by the front door, was now an autumn-red color. There were other colors beneath this one, and these, too, were memories. Times and tastes change.

New people came and they both agreed that the old radiators had to go. They weren't able to see the beauty beneath the many coats of paint. They couldn't hear the voices of memory, so they hired a contractor and he brought in his crew and they ripped out the old boiler, and the pipes, and the radiator, and they carted it all off to the dump.

And that old radiator, the one by the front door, faded in the sunlight and rusted in the rain, and waited patiently for nature to return its elements to the earth where it would join with those of a young man who had once tinkered and dreamed, and made something classic, something that served for a hundred years.

Something beautiful.

INDEX

www.ingramcontent.com/pod-product-compliance
Lightning Source LLC
Chambersburg PA
CBHW081505200326
41518CB00015B/2391